TECHNIQUES OF EXTENSION OF ANALYTIC OBJECTS

Lecture Notes in Pure and Applied Mathematics

1. *N. Jacobson,* Exceptional Lie Algebras
2. *L.-Å. Lindahl* and *F. Poulsen,* Thin Sets in Harmonic Analysis
3. *I. Satake,* Classification Theory of Semi-Simple Algebraic Groups
4. *F. Hirzebruch, W. D. Neumann,* and *S. S. Koh,* Differentiable Manifolds and Quadratic Forms
5. *I. Chavel,* Riemannian Symmetric Spaces of Rank One
6. *R. B. Burckel,* Characterization of C(X) among Its Subalgebras
7. *B. R. McDonald, A. R. Magid,* and *K. C. Smith,* Ring Theory: Proceedings of the Oklahoma Conference
8. *Yum-Tong Siu,* Techniques of Extension of Analytic Objects

Other volumes in preparation

TECHNIQUES OF EXTENSION OF ANALYTIC OBJECTS

Yum-Tong Siu

DEPARTMENT OF MATHEMATICS
YALE UNIVERSITY
NEW HAVEN, CONNECTICUT

MARCEL DEKKER, INC. New York 1974

MARCEL DEKKER, INC.
270 Madison Avenue, New York, New York 10016

LIBRARY OF CONGRESS CATALOG CARD NUMBER: 74-83962

ISBN: 0-8247-6168-5

Current printing (last digit):
10 9 8 7 6 5 4 3 2 1

PRINTED IN THE UNITED STATES OF AMERICA

To Sau-Fong and Brian

TABLE OF CONTENTS

PREFACE

This set of lecture notes is a reproduction (with minor modifications) of the notes I distributed to the students in the "troisième cycle" course "Techniques of Extension of Analytic Objects" I gave at the University of Paris VII in the second semester of 1971-1972 while I was on a leave of absence from Yale University.

The results presented here are known. However, some of the proofs (for example, the proof of the extension theorem for coherent analytic sheaves from a Hartogs' figure) are new. Most of the results appear here in book form for the first time.

The reader is assumed to have some familiarity with the basic theory of several complex variables, as given, for example, in Gunning-Rossi's "Analytic Functions of Several Complex Variables".

The appendices at the end of the chapters contain background material and are mainly for the convenience of the reader. Some of the material there are slightly different from the standard formulations usually found in the literature and some are special cases of well-known results for which direct proofs are supplied which are much easier than those of the general cases.

I wish to thank the University of Paris VII and Professor F. Norguet for their hospitality during my stay in

Paris. I would like also to thank the Alfred P. Sloan Foundation and the National Science Foundation for providing part of the financial support respectively during my leave of absence from Yale University and during periods in which this set of lecture notes was prepared and organized. Finally I wish to thank Betsy Buslovitz for her excellent typing.

TECHNIQUES OF EXTENSION OF ANALYTIC OBJECTS

INTRODUCTION

We are concerned with the theory of extension of what we call analytic objects. We will not define abstractly what an analytic object is. By an analytic object, we simply mean one of the following entities whose extension we will consider: holomorphic and meromorphic functions, analytic subvarieties, coherent analytic subsheaves and sheaves, holomorphic maps, etc. In this introduction, we will describe the main results, point out their interrelationships, and give a general idea of their proofs.

First, we want to say something about the domains on which the analytic objects are defined — the domains before extension and the domains after extension. We consider the following types:

$1)_n$ Hartogs' domains of order n

In $\mathbb{R}^2$ consider the following set

$$D' = \{0 \leq x < 1,\ a < y < 1\} \quad \{0 \leq x < \rho,\ 0 \leq y < 1\}$$

(where $0 \leq a < 1$ and $0 < \rho < 1$).

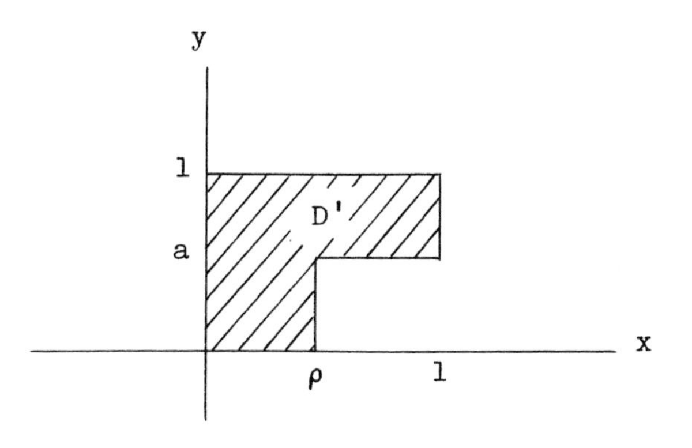

Consider the map

$$\varphi: \mathbb{C}^n \times \mathbb{C}^N \longrightarrow \mathbb{R}^2$$

defined by

$$\varphi(t_1, \ldots, t_n; z_1, \ldots, z_N) = (x,y)$$

where

$$x = \max_{1 \leq i \leq n} |t_i| ,$$

$$y = \max_{1 \leq j \leq N} |z_j| ,$$

$t_1, \ldots, t_n$ are the coordinates of $\mathbb{C}^n$ and $z_1, \ldots, z_N$ are the coordinates of $\mathbb{C}^N$. We call $D = \varphi^{-1}(D')$ a <u>Hartogs domain of order n</u>. Note that we have suppressed the number N, because, as we will see later, what determines the extendibility of an analytic object is the number n. From now on, when we talk of a Hartogs domain of order n, we assume implicitly that $N \geq 1$. The envelope of holomorphy of a Hartogs domain is a polydisc which we will call its <u>associated polydisc</u>. The first type of extension we want to consider is to extend an analytic object from a Hartogs domain to its associated polydisc.

$2)_n$ <u>Ring domain of order n</u>

In $\mathbb{R}^2$ we consider the set

$$D' = \{0 \leq x < 1, \; a < y < 1\}$$

(where $0 \leq a < 1$).

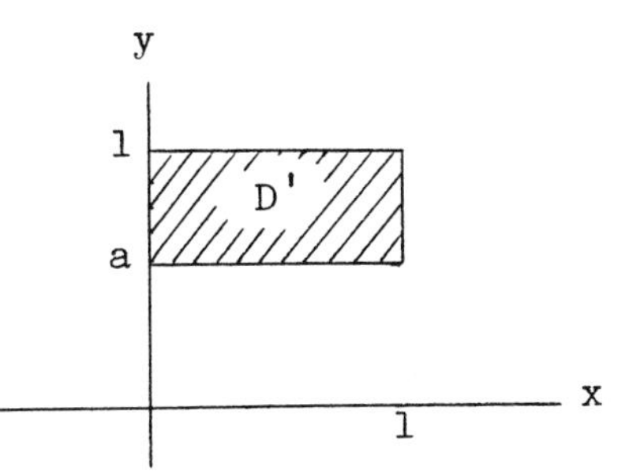

We call $D = \varphi^{-1}(D')$ a <u>ring domain of order n</u>, where φ is as in $1)_n$. The unit $(n+N)$-disc, which is the envelope of holomorphy when $N \geqq 2$, is called the <u>associated polydisc</u> of the ring domain. The second type of extension we want to consider is to extend an analytic object from a ring domain of order n (when $N \geqq 2$) to its associated polydisc.

$3)_n$ <u>Thullen type extension of order n</u>

Suppose Ω is an open subset of $\mathbb{C}^N$ and V is a subvariety of Ω of dimension n and G is an open subset of Ω which intersects every branch of V of dimension n. The extension of Thullen type is to extend an analytic object from $(\Omega - V) \cup G$ to Ω.

$4)_n$ <u>Extension across a subvariety of dimension n</u>

Suppose Ω is an open subset of $\mathbb{C}^N$ and V is a subvariety of dimension n in Ω. We want to extend an analytic object from $\Omega - V$ to Ω.

$5)_k$ <u>Extension across a closed subset with Hausdorff k-measure 0</u>

Suppose Ω is an open subset of $\mathbb{C}^N$ and A is a closed subset of Ω with Hausdorff k-measure 0 . We want to extend an analytic object from $\Omega - A$ to Ω . Recall here the definition of a Hausdorff measure. The Hausdorff k-measure of A , denoted by $h^k(A)$, is defined as

$h^k(A) = \sup_{\epsilon > 0} h^k_\epsilon(A)$, where

$$h^k_\epsilon(A) = \inf\left\{ \sum_{i=1}^{\infty} (\text{diameter of } A_i)^k \,\middle|\, A \subset \bigcup_{i=1}^{\infty} A_i, \right.$$

$$\left. \text{diameter of } A_i < \epsilon \right\}.$$

Note that a subvariety of dimension n has Hausdorff (2n+1)-measure 0 .

6) Extension across $\mathbb{R}^n$

Suppose Ω is an open subset of $\mathbb{C}^n$. We want to extend an analytic object from $\Omega - \mathbb{R}^n$ to Ω .

Besides the above six types of extension, we also want to consider a type which is closely related to and is more general than the Hartogs domain type.

$0)_n$ Ring domain of order n with extension already along a thick set of fibres

Suppose D is the ring domain of order n given in $2)_n$. For every t belonging to the unit n-disc Δ^n , let

$$F_t = \{(t, z_1, \ldots, z_N) \in \mathbb{C}^{n+N} \mid a < \max_{1 \leq j \leq N} |z_j| < 1\} ,$$

$$\tilde{F}_t = \{(t, z_1, \ldots, z_N) \in \mathbb{C}^{n+N} \mid \max_{1 \leq j \leq N} |z_j| < 1\} .$$

A subset A of an open subset G of $\mathbb{C}^n$ is called a <u>thin</u> set if $A \subset \bigcup_{i=1}^{\infty} A_i$ and A_i is a subvariety of codimension ≥ 1 in some open subset of G. A subset of G is called <u>thick</u> if it is not thin. Suppose we have an analytic object X defined on D and we have a thick set $A \subset \Delta^n$ and suppose, for every $t \in A$, the "analytic restriction" of X to F_t can be extended to $\tilde{F}_t$. Under these conditions, we want to extend X to the associated polydisc of D. For this type of extension, we assume only that $N \geq 1$.

The above types of extension are related as follows (for $n \geq 1$):

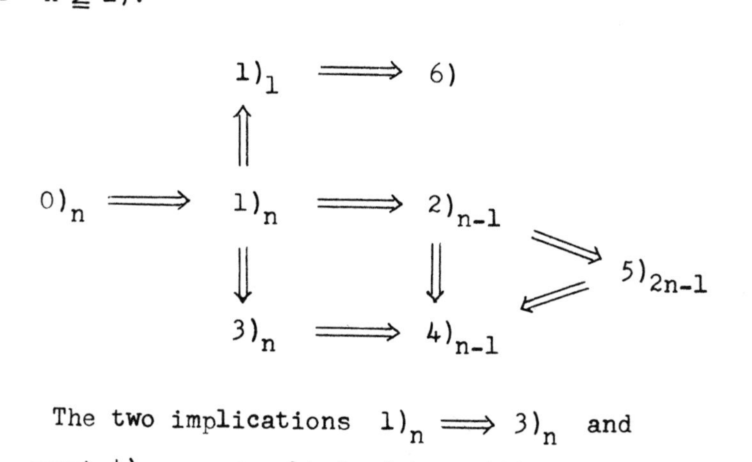

The two implications $1)_n \Longrightarrow 3)_n$ and $2)_{n-1} \Longrightarrow 4)_{n-1}$ are obtained by setting $a = 0$ in $1)_n$ and $2)_{n-1}$.

The implication $3)_n \Longrightarrow 4)_{n-1}$ follows from the fact that a subvariety of dimension $n - 1$ is locally contained in a subvariety of dimension n.

The implication $1)_n \Longrightarrow 2)_{n-1}$ is obtained by considering the following figure which represents the ring domain of order n :

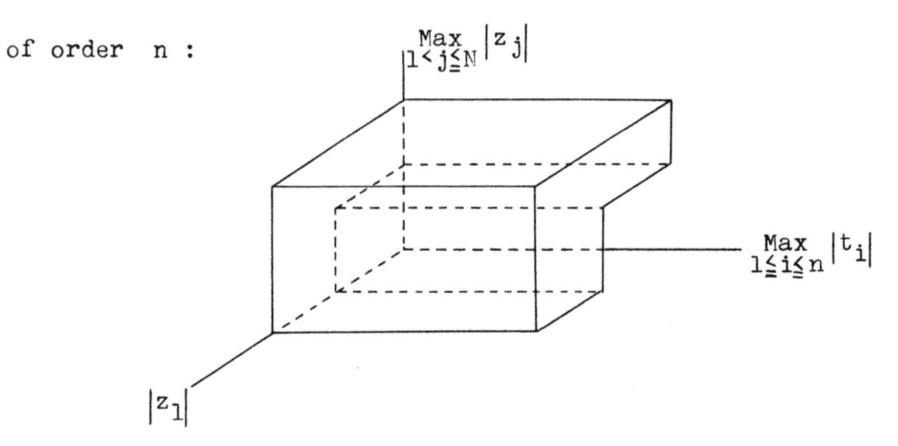

For the implication $1)_1 \Longrightarrow 6)$, we consider the map

$$(z_1, \ldots, z_n) \longmapsto (e^{iz_1}, \ldots, e^{iz_n}) .$$

It maps $\mathbb{R}^n$ to C^n , where C is the unit circle in $\mathbb{C}$.
Let

$$\varphi(z_1, \ldots, z_n) = |z_1|^2 + \ldots + |z_n|^2 - n .$$

C^n lies in $\{\varphi = 0\}$. For every point $P \in C^n$ we can find a Hartogs domain D of order 1 (in some local coordinates system) such that $D \subset \{\varphi > 0\}$ and P belongs to the associated polydisc of D .

For the implication $2)_{n-1} \Longrightarrow 5)_{2n-1}$ we have to make use of a lemma due to Federer to find a (complex) hyperplane H of codimension $n - 1$ such that $h^1(H \cap A) = 0$. We can assume that

$$H = \{z_1 = \ldots = z_{n-1} = 0\} .$$

Since $h^1(H \cap A) = 0$, after a change of coordinates involving only $z_n, \ldots, z_N$, we have

$$H \cap A \cap \{\max_{n \leq j \leq N} |z_j| = 1\} = \emptyset .$$

For $\epsilon > 0$ small enough, A is disjoint from

$$\left\{ \max_{1 \leqq i \leqq n-1} |z_i| < \epsilon, \ 1-\epsilon < \max_{n \leqq j \leqq N} |z_j| < 1+\epsilon \right\}$$

which is a ring domain of order $n - 1$.

The other implications in the diagram are obvious.

We would like to remark that $1)_n$ is equivalent to the following type of extension:

$1)_n'$ <u>Extension across strictly n-pseudoconcave boundaries</u>

Suppose φ is a real-valued C^2 function on an open subset of $\mathbb{C}^\ell$. φ is called a <u>strictly</u> n-<u>pseudoconvex</u> function if at every point the hermitian matrix

$$\left(\frac{\partial^2 \varphi}{\partial z_i \, \partial \bar{z}_j} \right)_{1 \leqq i, \, j \leqq \ell}$$

(where $z_1, \ldots, z_\ell$ are the coordinates of $\mathbb{C}^\ell$) has at least $\ell - n + 1$ positive eigenvalues. A domain Ω in $\mathbb{C}^N$ is said to be <u>strictly</u> n-<u>pseudoconcave</u> at a boundary point P if there exist an open neighborhood U of P and a strictly n-pseudoconvex function ρ on U such that

$$\Omega \cap U = \{x \in U \,|\, \rho(x) > 0\} .$$

$1)_n'$ means that an analytic object is defined on Ω and we want to extend it to $W \cup \Omega$ for some open neighborhood W of P.

We have $1)_n \Longrightarrow 1)_n'$, because we can find a Hartogs domain D of order n (in some local coordinates system) contained in Ω whose associated polydisc contains P. We have $1)_n' \Longrightarrow 1)_n$, because we can exhaust the associated polydisc $\tilde{D}$ of a Hartogs domain D of order n by an increasing continuous family of subdomains U_α $(0 < \alpha < 1)$ of $\tilde{D}$ (cf. [28, (10.2)]) such that

i) $U_\alpha \subset D$ when α is small,

ii) $\partial U_\alpha \cap (\tilde{D}-D)$ is compact,

iii) U_α is strictly n-pseudoconcave at every point of $\partial U_\alpha \cap (\tilde{D}-D)$.

We now state the results of extension of analytic objects and point out their interrelationships.

I. Holomorphic and meromorphic functions

$0)_n$ is true. This result is due to E. E. Levi [11]. There is a deeper theorem of extension due to Rothstein [15]. Its statement is as follows:

> Suppose f is a meromorphic (holomorphic) function on the unit $(n+1)$-disc and suppose, for every t belonging to the unit n-disc, $f_t(z) = f(t,z)$ can be extended meromorphically (holomorphically) to $\{|z| < 2\} \subset \mathbb{C}$. Then f can be extended meromorphically (holomorphically) to $\{|t_1| < 1, \ldots, |t_n| < 1, |z| < 2\} \subset \mathbb{C}^{n+1}$.

Levi's theorem depends only on Laurent expansion. Rothstein's theorem depends on the following property of subharmonic functions:

Let $D = \bigcup_{\nu} D_{\nu}$ be an open subset of the unit 1-disc Δ, where D_{ν} is connected and open and (diameter D_{ν}) $\geqq \delta > 0$ for all ν. Let $z_0 \in \Delta \cap \partial D$. If u is a subharmonic function on Δ, then $u(z_0) \leqq \sup_{z \in D} u(z)$.

II. Subvarieties

a) $0)_n$ is true for subvarieties whose every branch has dimension $\geqq n+1$. Here the analytic restriction to a fiber means the intersection of the subvariety with the fiber.

$1)_n$ is due to Rothstein [16].

The proof uses projections, special analytic polyhedra, analytic covers, elementary symmetric polynomials, and the extension of holomorphic functions.

b) $3)_n$ is true for subvarieties whose every branch has dimension $\geqq n$. This is the theorem of Thullen-Remmert-Stein. The special case of $N = 1$ is due to Thullen [32]. The general case is due to Remmert-Stein [14]. The proof of the general case is reduced to the special case by suitable projections and the special case is proved by using the following theorem of Radó:

If f is a continuous function on the unit 1-disc and is holomorphic whenever it is nonzero, then it is holomorphic on the whole disc.

$5)_{n-1}$ is true for subvarieties whose every branch has

dimension $\geq n$. This is due to Shiffman [19]. This is not as deep as the Thullen-Remmert-Stein theorem. The proof uses projections and function extensions.

c) $4)_n$ is true for subvarieties of pure dimension k whose Hausdorff $2k$-measure is finite. This is due to Bishop [3]. It is a deep theorem. The proof depends on the estimation of certain curve-lengths, which is in turn dependent on the Poisson kernel.

d) 6) is true for <u>self-conjugate</u> subvarieties of pure dimension 1 , i.e. subvarieties V satisfying

$$(z_1, \ldots, z_n) \in V \implies (\bar{z}_1, \ldots, \bar{z}_n) \in V .$$

The proof makes use of quadratic projections and Schwarz reflection principle. This result is due to Alexander [1], Becker [2].

III. <u>Coherent subsheaves</u>

$0)_n$ is true for coherent subsheaves $\mathscr{A}$ satisfying the relative gap-sheaf condition $\mathscr{A}_{[n]} = \mathscr{A}$. The gap-sheaf condition means the following: for a subsheaf $\mathscr{A}$ of a sheaf $\mathscr{J}$, define $\mathscr{A}_{[n]}$, called the nth <u>relative gap-sheaf</u> of $\mathscr{A}$ in $\mathscr{J}$, as the subsheaf of $\mathscr{J}$ whose stalk at x consists of all $s \in \mathscr{J}_x$ such that there exist an open neighborhood U of x , $t \in \Gamma(U,\mathscr{J})$, and a subvariety A of U of dimension $\leq n$ with $s = t_x$ and $t_y \in \mathscr{A}_y$ for $y \in U-A$. It is always true that $\mathscr{A} \subset \mathscr{A}_{[n]}$. $\mathscr{A}_{[n]} = \mathscr{A}$ means that the two subsheaves coincide. In the extension of type $0)_n$, $\mathscr{J}$ is a coherent sheaf defined on the associated polydisc and $\mathcal{A}$ is a coherent subsheaf of $\mathcal{J}$ defined on the ring domain. The analytic restriction of $\mathcal{A}$ to a fiber F means the image of $\mathcal{A} \otimes \mathcal{O}_F \longrightarrow \mathcal{J} \otimes \mathcal{O}_F$, where $\mathcal{O}_F$ is the structure sheaf of F. This theory is due to Thimm [30,31], Siu-Trautmann [27].

Observe that, if V is a subvariety whose every branch has dimension $\geq n+1$, then the ideal-sheaf $\mathcal{J}_V$ of V as a subsheaf of the structure sheaf of $\mathbb{C}^{n+N}$ satisfies $(\mathcal{J}_V)_{[n]} = \mathcal{J}_V$. Hence $0)_n$ for subsheaves generalizes Rothstein's theorem for subvarieties.

The proof of $0)_n$ for subsheaves depends on reducing the general case via projection to a special case whose extendibility follows from the extendibility of meromorphic functions. In the reduction process by projection, the extendibility of subvarieties has to be used. So essentially:

subsheaf extension = subvariety extension + meromorphic function extension.

IV. Coherent Sheaves

$0)_n$ is true for coherent sheaves $\mathcal{F}$ satisfying the gap-sheaf condition $\mathcal{F}^{[n]} = \mathcal{F}$. Here $\mathcal{F}^{[n]}$, called the nth absolute gap-sheaf of $\mathcal{F}$, is the sheaf associated to the following presheaf

$$U \longmapsto \mathcal{F}^{[n]}(U) = \operatorname{ind.}\lim_{A} \Gamma(U-A, \mathcal{F})$$

where A runs through the set of all subvarieties of U of

dimension $\leqq n$. Always we have a canonical map $\mathcal{F} \longrightarrow \mathcal{F}^{[n]}$. $\mathcal{F}^{[n]} = \mathcal{F}$ means that this canonical map is an isomorphism. When $n = 1$, $2)_{n-1}$ is due to Trautmann [33]. Siu [22,24] used Grauert's power series method to obtain $2)_n$ and $0)_n$; and Frisch-Guenot [5] provided an elegant proof of $4)_{n-1}$ by Douady's method of privileged sets. In these lecture notes, we will give a new proof of $0)_n$ which uses neither the series method nor the method of privileged sets. The idea is to use projections to reduce it to a special case and for this special case the method of power series and the method of privileged sets are unnecessary. This proof is more in line with the proofs used in the extension of functions, subvarieties, and subsheaves. $0)_n$ for sheaves generalizes $0)_n$ for subsheaves.

V. Holomorphic Maps

A holomoprhic map from a punctured ball of dimension ≥ 2 to a compact Kähler manifold can be extended to a meromorphic map defined on the whole ball. This depends on Bishop's theorem applied to the graph and also on the fact that a closed nonnegative $(1,1)$ current on the punctured ball can be extended to the whole ball. This theory is due to Griffiths [6] and Shiffman [20].

CHAPTER 1

EXTENSION OF HOLOMORPHIC AND MEROMORPHIC FUNCTIONS

The main results in this Chapter are a theorem of E. E. Levi and a theorem of W. Rothstein.

Let $\Delta(r) = \{z \in \mathbb{C} \mid |z| < r\}$ and $\Delta = \Delta(1)$.

(1.1) Theorem (Levi). $0)_n$ is true for meromorphic functions, i. e. Suppose $f(z_1, \ldots, z_n, w)$ is a meromorphic function on $\Omega \times (\Delta(r) - \bar{\Delta})$, where $r > 1$ and $\Omega \subset \mathbb{C}^n$ is open and connected. Suppose A is a thick set in Ω and suppose, for every $(z_1, \ldots, z_n) \in A$, $f(z_1, \ldots, z_n, w)$ extends to a meromorphic function on $\Delta(r)$. Then f extends to a meromorphic function on $\Omega \times \Delta(r)$.

Proof. 1° Special case. Assume f is holomorphic on $\Omega \times (\Delta(r) - \bar{\Delta})$. We have a Laurent series expansion

$$f(z,w) = \sum_{\nu=-\infty}^{\infty} c_\nu(z) w^\nu \qquad (1 < |w| < r)$$

where $c_\nu(z)$ is holomorphic on Ω. Let A_p be the set of all $z \in A$ such that the meromorphic extension of $f(z,w)$ on $\Delta(r)$ has at most p poles (with multiplicities counted). Since $A = \bigcup_{p=0}^{\infty} A_p$, there exists some p such that A_p is thick. Consider the vectors

$$V_\nu = (c_{-\nu}, c_{-\nu-1}, \ldots, c_{-\nu-p}) \qquad (\nu \geq 1)$$

over the field F of meromorphic functions on Ω. We claim that the vector subspace $\mathcal{V} \subset F^{p+1}$ spanned by $\{V_\nu\}_{\nu=1}^{\infty}$ has dimension $\leq p$ over F, i.e.

$$D: = \det \begin{pmatrix} c_{-\nu_1} & c_{-\nu_1-1} & \cdots & c_{-\nu_1-p} \\ c_{-\nu_2} & c_{-\nu_2-1} & \cdots & c_{-\nu_2-p} \\ & & & \\ c_{-\nu_{p+1}} & c_{-\nu_{p+1}-1} & \cdots & c_{-\nu_{p+1}-p} \end{pmatrix} \equiv 0 \quad \text{on } \Omega$$

for all $1 \leq \nu_1 < \ldots < \nu_{p+1} < \infty$. We know that, for every $z \in A_p$, there exist $a_0, \ldots, a_p \in \mathbb{C}$, not all zero, such that

$$\left(\sum_{\mu=0}^{p} a_\mu w^\mu\right)\left(\sum_{\nu=-\infty}^{\infty} c_\nu(z) w^\nu\right) \quad \text{is holomorphic on } \{|w| < r\} ,$$

i.e.

$$a_0 c_{-\nu}(z) + a_1 c_{-\nu-1}(z) + \ldots + a_p c_{-\nu-p}(z) = 0 \quad \text{for } \nu \geq 1 .$$

Hence $D = 0$ at every point of A_p. Since A_p is thick, $D \equiv 0$ on Ω. Let $V_{\nu_1}, \ldots, V_{\nu_k}$ span $\mathcal{V}$, where $k \leq p$. Then, for every ν, there exist holomorphic functions $\alpha^{(\nu)}, \alpha_1^{(\nu)}, \ldots, \alpha_k^{(\nu)}$ holomorphic on Ω with $\alpha^{(\nu)} \not\equiv 0$ such that

$$(*) \qquad \alpha^{(\nu)} V_\nu = \sum_{i=1}^{k} \alpha_i^{(\nu)} V_{\nu_i} .$$

Since $k \leq p$, there exist holomorphic functions $\beta_0, \ldots, \beta_p$ on Ω, not all identically zero, such that

$$\sum_{\mu=0}^{p} \beta_\mu c_{-\nu_i-\mu} \equiv 0 \qquad \text{on } \Omega \text{ for } 1 \leqq i \leqq k .$$

By virtue of (*),

$$\sum_{\mu=0}^{p} \beta_\mu c_{-\nu-\mu} = 0 \qquad \text{on } \Omega \text{ for } 1 \leqq \nu < \infty .$$

Hence

$$g: = \Big(\sum_{\mu=0}^{p} \beta_\mu(z)w^\mu\Big)\Big(\sum_{\nu=-\infty}^{\infty} c_\nu(z)w^\nu\Big)$$

is holomorphic on $\Omega \times \Delta(r)$. The meromorphic function $\Big(\sum_{\mu=0}^{p} \beta_\mu(z)w^\mu\Big)^{-1} g$ extends f .

2° General case. Let A' be the set of all $z \in A$ such that, for every open neighborhood U of z , $U \cap A$ is thick. It is easy to see that A' is thick. Fix arbitrarily $z^0 \in A'$ such that $\{z^0\} \times (\Delta(r) - \bar{\Delta})$ is not contained in the pole-set P of f. Such a z^0 exists, otherwise P contains $A' \times (\Delta(r) - \bar{\Delta})$ which is thick in $\Omega \times (\Delta(r) - \bar{\Delta})$ (cf. (2.A.8)). There exists an open neighborhood U of z^0 and $1 \leqq \alpha < \beta \leqq r$ such that f is holomorphic on $U \times (\Delta(\beta) - \overline{\Delta(\alpha)})$. By 1° , f can be extended meromorphically to $U \times \Delta(r)$. Let Ω' be the set of all $z \in \Omega$ such that, for some open neighborhood W of z in Ω , f can be extended meromorphically to $W \times \Delta(r)$. $\Omega' \neq \emptyset$, because $z^0 \in \Omega'$. By using coordinates transformations of the form

$$\begin{cases} z_j' = z_j + \epsilon w \\ w = w \end{cases} \qquad (\epsilon > 0)$$

and the preceding argument, we conclude easily that Ω' is closed in Ω. Hence $\Omega = \Omega'$. Q.E.D.

For holomorphic functions, by Laurent series expansions, it is easy to see that the following stronger theorem holds:

(1.2) Theorem. *Suppose $f(z,w)$ is a holomorphic function on $\Omega \times (\Delta(r) - \bar{\Delta})$, where $r > 1$ and $\Omega \subset \mathbb{C}^n$ is open and connected. Suppose $A \subset \Omega$ is not contained in any subvariety of Ω of codimension ≥ 1. Suppose, for every $z \in A$, $f(z,w)$ extends to a holomorphic function on $\Delta(r)$. Then f extends to a holomorphic function on $\Omega \times \Delta(r)$.*

(1.3) Remark. In (1.1) we cannot replace the thickness of A by the condition that A is not contained in any subvariety of Ω of codimension ≥ 1. The following is a trivial counter-example: $\Omega = \mathbb{C}$. $A = \{ \frac{1}{\nu} \mid \nu$ is a positive integer$\}$. For $\nu \geq 1$ let $c_\nu(z)$ be a holomorphic function on such that $|c_\nu(z)| \leq \frac{1}{\nu!}$ for $|z| \leq \nu$ and such that the zero-set of $c_\nu(z)$ is precisely $\{\frac{1}{1}, \frac{1}{2}, \cdots, \frac{1}{\nu}\}$. Let

$f(z,w) = \sum_{\nu=1}^{\infty} c_\nu(z) w^{-\nu}$. f is holomorphic on $\mathbb{C} \times (\mathbb{C} - \{0\})$ and, for $z \in A$, $f(z,w)$ is a meromorphic function on $\mathbb{C}$. However, f is not a meromorphic function on $\mathbb{C} \times \mathbb{C}$.

To get Rothstein's theorem, we have to have the subharmonicity of $-\log$ of the radius of meromorphy and the lemma on subharmonic functions stated on p. 9. So we will first define the radius of meromorphy and prove the subharmonicity of its $-\log$. Then, to prove the lemma on subharmonic functions (see (1.9)), we will make use of the $\frac{1}{4}$ -theorem of Koebe-Bieberbach (see (1.A.3) of the Appendix of Chapter 1).

(1.4) Proposition. *Suppose Ω is an open subset of $\mathbb{C}$ and D is a connected open neighborhood of 0 in $\mathbb{C}$. Suppose f is a meromorphic function on $\Omega \times D$. For $z \in \Omega$ define the radius of meromorphy $r(z)$ of f to be the supremum of all ρ such that f can be extended meromorphically to an open neighborhood of $\{z\} \times \Delta(\rho)$. Then $-\log r(z)$ is subharmonic on Ω.*

Proof. Let K be a closed disc contained in Ω. Let h be a harmonic function defined on an open neighborhood of K such that $-\log r(z) \leq h(z)$ for $z \in \partial K$. We need only show that $-\log r(z) \leq h(z)$ for $z \in K$. Without loss of generality we can assume that $K = \overline{\Delta}$. $h = \operatorname{Re} f$ for some holomorphic function f defined on an open neighborhood of $\overline{\Delta}$.

Define $\varphi: \mathbb{C}^2 \longrightarrow \mathbb{C}^2$ by $\varphi(z,\zeta) = (z,w)$, where $w = \zeta e^{-f(z)}$. φ is a biholomorphic map. For $|z| = 1$ and $|\zeta| < 1$, we have

$$|w| = |\zeta| e^{-h(z)} < e^{-h(z)} \leq r(z).$$

Hence $f \circ \varphi$ can be extended meromorphically to an open neighborhood of $\{|z| = 1, |\zeta| < 1\}$. There exists $\epsilon > 0$ such that φ maps $\{|z| \leq 1, |\zeta| < \epsilon\}$ into $\Omega \times D$. $f \circ \varphi$ is defined and meromorphic on $\{|z| \leq 1, |\zeta| < \epsilon\}$. By (1.1) (more precisely, by $1)_1$), $f \circ \varphi$ can be extended meromorphically to an open neighborhood of $\{|z| \leq 1, |\zeta| < 1\}$. Since the image of $\{|z| \leq 1, |\zeta| < 1\}$ under φ is $\{|z| \leq 1, |w| < e^{-h(z)}\}$, f can be extended meromorphically to $\{|z| \leq 1, |w| < e^{-h(z)}\}$. Hence $-\log r(z) \leq h(z)$ for $|z| \leq 1$. Q.E.D.

(1.5) Remark. In (1.4), when Ω is an open subset of $\mathbb{C}^n$ instead of $\mathbb{C}$, the same argument shows that $-\log r(z)$ is plurisubharmonic on Ω. When D is a neighborhood of 0 in $\mathbb{C}^n$ instead of in $\mathbb{C}$, we can define $r(z)$ as the supremum of all ρ such that f can be extended meromorphically to an open neighborhood of $\{z\} \times (\Delta(\rho))^n$. The same conclusion holds. The only modification in the argument is to define $\varphi(z, \zeta_1, \ldots, \zeta_n) = (z, w_1, \ldots, w_n)$ with $w_\mu = \zeta_\mu e^{-f(z)}$. (1.4) holds also for analytic objects other than meromorphic functions as long as they satisfy the extension property $1)_1$ (or $1)_n$ when D is a neighborhood of 0 in $\mathbb{C}^n$) and the "Identity Theorem".

(1.6) Lemma (E. Schmidt [18]). *Let C be a Jordan curve in the closed unit disc $\overline{\Delta}$ joining 0 to 1 such that $C \cap \partial\Delta = \{1\}$. Let $G = \Delta - C$. Suppose f maps G biholomorphically onto the open upper half H of Δ such that $\partial\Delta$ corresponds to $[-1,1]$ and C corresponds to the upper half unit circle. Let $a \in G$ such that $f(a) = it_0$ with $t_0 > 0$. Then* $t_0 \geq \dfrac{1 - \sqrt{|a|}}{1 + \sqrt{|a|}}$.

Proof. Let $\zeta = f(z)$ and let $z = g(\zeta)$ be the inverse of f. Since g is nowhere zero on H, a branch of $\log g$ exists on H. It extends continuously to $(-1,1)$.

$$\log g(it_0) - \log g(0) = \int_{t=0}^{t=t_0} \frac{1}{z}\left(\frac{dz}{d\zeta}\right)_{\zeta=it} dt .$$

$$\log \frac{1}{|a|} = \left|\mathrm{Re}\left(\log g(it_0) - \log g(0)\right)\right|$$

$$\leq |\log g(it_0) - \log g(0)| \leq \int_{t=0}^{t=t_0} \frac{1}{|z|} \left|\frac{dz}{d\zeta}\right|_{\zeta=it} dt.$$

Now we have to use the $\frac{1}{4}$-theorem to estimate $\frac{1}{|z|}\left|\frac{dz}{d\zeta}\right|_{\zeta=it}$.

Let

$$\zeta' = h(\zeta) = \frac{\zeta - it}{1 + it\zeta} .$$

Since

$$\left(\frac{d\zeta'}{d\zeta}\right)_{\zeta=it} = \frac{1}{1-t^2},$$

we have

$$\left|\frac{dz}{d\zeta}\right|_{\zeta=it} = \frac{1}{1-t^2}\left|\frac{dz}{d\zeta'}\right|_{\zeta'=0}$$

By Schwarz reflection principle, g can be extended to a biholomorphic map

$$\tilde{g}: \Delta \longrightarrow \mathbb{C} - (C \cup C'),$$

where C' is the reflection of C with respect to $\partial\Delta$.

Let

$$\alpha = \left(\frac{dz}{d\zeta'}\right)_{\zeta'=0} = (\tilde{g}h^{-1})'(0) .$$

Define

$$\varphi = \frac{1}{\alpha}\left(\tilde{g}h^{-1} - g(it)\right) .$$

φ is a univalent function Δ with $\varphi(0) = 0$ and $\varphi'(0) = 1$. By the $\frac{1}{4}$-theorem of Koebe-Bieberbach (see (1.A.3) of the Appendix of Chap. 1), we have $\varphi(\Delta) \supset \Delta(\frac{1}{4})$. Since $\tilde{g}$ is never zero, the point $-\frac{1}{\alpha}g(it)$ is not in the image of φ . Hence

$$\left|-\frac{1}{\alpha}g(it)\right| \geqq \frac{1}{4} ,$$

i.e.

$$\left|\frac{dz}{d\zeta'}\right|_{\zeta'=0} \leqq 4|g(it)| .$$

It follows that

$$\frac{1}{|z|}\left|\frac{dz}{d\zeta}\right|_{\zeta=it} \leqq \frac{4}{1 - t^2} .$$

Consequently,

$$\log\frac{1}{|a|} \leqq \int_{t=0}^{t=t_0} \frac{4}{1 - t^2}\, dt = 2\log\frac{1 + t_0}{1 - t_0}$$

and

$$t_0 \geqq \frac{1 - \sqrt{|a|}}{1 + \sqrt{|a|}} .$$

Q.E.D.

(1.7) Lemma. Suppose C and G are as in (1.6). Suppose $\omega(z)$ is the harmonic measure of C with respect to G, i.e. ω is harmonic on G and approaches 1 on C and 0 at other points of G. Then, for every $\epsilon > 0$, there exists $\delta = \delta(\epsilon) > 0$, independent of C, such that $\omega(z) > 1-\epsilon$ for $|z| < \delta$.

Proof. Choose a biholomorphic map $\tau: G \longrightarrow H$ (where H is as in (1.6)) such that $\partial\Delta$ corresponds to $[-1,1]$ and C corresponds to the upper half unit circle. This is possible, because we can map both G and H biholomorphically onto Δ and find a fractional linear transformation mapping any prescribed triple of distinct points to another prescribed triple of distinct points.

Take $a \in G$. Let γ be the circle which is perpendicular to both $\mathbb{R}$ and $\partial\Delta$ and which passes through $\tau(a)$. Let b be the point where γ intersects the upper half of $\partial\Delta$. Let θ be the fractional linear transformation mapping the triple of $(-1,b,1)$ to the triple $(-1,i,1)$. Let $\sigma_a = \theta\circ\tau$. Then σ_A is a biholomorphic map from G to H such that $\sigma_a(a) = it_a$ with $t_a > 0$ and such that $\partial\Delta$ corresponds to $[-1,1]$ and C corresponds to the upper half of $\partial\Delta$. By (1.6),

$$t_a \geq \frac{1 - \sqrt{|a|}}{1 + \sqrt{|a|}} .$$

Let ω' be the harmonic measure of the upper half unit circle with respect to H. Then $\omega\circ\sigma_a^{-1} = \omega'$. We have

$$\omega(a) = \omega'(it_a) = 1 - (1 - \omega'(it_a))$$

$$\geq 1 - \sup\left\{1 - \omega'(it) \,\middle|\, \frac{1-\sqrt{|a|}}{1+\sqrt{|a|}} \leq t \leq 1\right\}.$$

Since

$$\sup\left\{1 - \omega'(it) \,\middle|\, \frac{1-\sqrt{|a|}}{1+\sqrt{|a|}} \leq t \leq 1\right\}$$

approaches 0 as $|a| \longrightarrow 0$, the lemma follows. Q.E.D.

(1.8) Lemma. *Suppose* u *is a subharmonic function on* Δ, *and suppose* $d > 0$ *and* $a, a_\nu, b_\nu \in \Delta$ $(1 \leq \nu < \infty)$ *such that* $a_\nu \longrightarrow a$ *and* $|a_\nu - b_\nu| \geq d$. *Let* C_ν *be a Jordan curve in* Δ *joining* a_ν *to* b_ν. *If* $u \leq A$ *for some* A *on all* C_ν, *then* $u(a) \leq A$.

Proof. Choose $\nu_0 \geq 1$, $0 < e < d$, and $|a| < \rho < 1$, such that

$$\{|z-a_\nu| \leq e\} \subset \Delta(\rho)$$

for $\nu \geq \nu_0$. Let C_ν' be the portion of C_ν from a_ν to the first point of intersection of C_ν with $\{|z-a_\nu| = e\}$. By (1.7), for every $\epsilon > 0$, there exists $\delta = \delta(\epsilon) > 0$ such that $\omega_\nu(z) > 1-\epsilon$ whenever $|z-a_\nu| < \delta$, where ω_ν is the harmonic measure of C_ν' with respect to

$$\{|z-a_\nu| < e\} - C_\nu'.$$

Let $M = \sup_{z \in \overline{\Delta(\rho)}} u(z)$. If $A \geq M$, there is nothing to prove. So we assume that $A < M$. Since

$$M - u \geqq (M-A)\,\omega_\nu$$

on $\{|z-a_\nu| < e\} - C_\nu'$ for $\nu \geqq \nu_0$, we have

$$M - u > (M-A)(1-\epsilon)$$

on $\{|z-a_\nu| < \delta\}$ for $\nu \geqq \nu_0$. Since $|a-a_\nu| < \delta$ for ν sufficiently large, it follows that $M - u(a) > (M-A)(1-\epsilon)$ for every $\epsilon > 0$. Hence $u(a) \leqq A$. Q.E.D.

(1.9) Lemma. *Let* $d > 0$. *Suppose* D_ν $(\nu \in I)$ *is a connected open subset of* Δ *such that the diameter of* D_ν *is* $\geqq d$ *for all* $\nu \in I$. *Let* $D = \bigcup_{\nu \in I} D_\nu$ *and* $z_0 \in \Delta \cap \partial D$. *If* u *is a subharmonic function on* Δ *and* $u \leqq A$ *on* D, *then* $u(z_0) \leqq A$.

Proof. There exists $a_k \in D$ $(1 \leqq k < \infty)$ such that $a_k \longrightarrow z_0$ as $k \longrightarrow \infty$. a_k is in some D_{ν_k}. There exists $b_k \in D_{\nu_k}$ such that $|a_k-b_k| \geqq \frac{d}{3}$ (otherwise the diameter of D_{ν_k} is $< d$). Join a_k to b_k by a Jordan curve C_k in D_{ν_k}. The lemma follows from (1.8). Q.E.D.

For a closed subset A of an open subset Ω of $\mathbb{C}^n$, we denote by $E_\Omega(A)$ the union of A and all components of $\Omega - A$ which are relatively compact in Ω.

(1.10) Lemma. *Let* $N \subset \Delta$ *and* $M = E_\Delta(\overline{N} \cap \Delta)$. *Let* U *be a connected open neighborhood of* Δ *in* $\mathbb{C}$. *Suppose*

$f\colon M \times U \longrightarrow \mathbb{C}$ _such that_

(i) f _is continuous on_ $M \times \Delta$ _and holomorphic on_ $M^0 \times \Delta$ (_where_ M^0 = _the interior of_ M), _and_

(ii) $f_z = f(z,\cdot)$ _is holomorphic on_ U _and_ $\|f_z\|_U \leq A < \infty$ _for some fixed_ A _and for all_ $z \in N$ (_where_ $\| \ \|_U$ _denotes the_ sup _norm on_ U).

Then f _is continuous on_ $M \times U$ _and holomorphic on_ $M^0 \times U$.

Proof. Obviously we need only consider the case $U = \Delta(r)$ with $r > 1$. Expand f in power series:

$$f(z,w) = \sum_{\nu=0}^{\infty} a_\nu(z) w^\nu \qquad (w \in \Delta),$$

where a_ν is continuous on M and holomorphic on M^0. Since $\|f_z\|_{\Delta(r)} \leq A$, $|a_\nu(z)| \leq Ar^{-\nu}$ for $z \in N$. By the maximum modulus principle, $|a_\nu(z)| \leq Ar^{-\nu}$ for $z \in M$. The lemma follows. Q.E.D.

(1.11) __Lemma.__ _Let_ V_1, V_2 _be connected open subsets of_ $\mathbb{C}$ _such that_ $\Delta \subset V_1 \cup V_2$. _Let_ $f\colon \Delta \times (V_1 \cup V_2) \longrightarrow \mathbb{C} \cup \{\infty\}$ _be such that_ f _is nowhere zero holomorphic on_ $\Delta \times \Delta$ _and, for all_ $z \in \Delta$, $f_z = f(z,\cdot)$ _is meromorphic on_ $V_1 \cup V_2$. _Let_ $A < \infty$ _and let_ N _be the set of all_ $z \in \Delta$ _such that_ $\|f_z\|_{V_1} \leq A$ _and_ $\left\|\frac{1}{f_z}\right\|_{V_2} \leq A$. _Then_

(i) N _is closed in_ Δ _and_ $E_\Delta(N) = N$, _and_

(ii) f _is holomorphic on_ $N^0 \times V_1$ _and_ $\frac{1}{f}$ _is holomorphic on_ $N^0 \times V_2$.

Proof. Follows readily from (1.10). Q.E.D.

(1.12) _Theorem_ (_Rothstein_). _Let_ $r > 1$. _Suppose_ f _is a meromorphic function on_ $\Delta \times \Delta$ _such that_ $f_z = f(z,\cdot)$ _extends meromorphically to_ $\Delta(r)$ _for all_ $z \in \Delta$. _Then_ f _extends meromorphically to_ $\Delta \times \Delta(r)$.

Proof. It is easy to see that without loss of generality we can assume that f is nowhere zero holomorphic on $\Delta \times \Delta$. Let $r(z)$ be the radius of meromorphy of f , i.e. $r(z)$ equals to the sup of all ρ such that f can be extended meromorphically to an open subset of $\{z\} \times \Delta(\rho)$. Let $u(z) = -\log r(z)$. By (1.4), u is subharmonic. Fix arbitrarily $1 < q < r$. Let G be the set of all $z \in \Delta$ such that, for some open neighborhood W of z in Δ , f can be extended meromorphically to $W \times \Delta(q)$. We are going to show that $G = \Delta$.

First, we show that G is dense in Δ . Introduce the following notation: Let $A > 0$, $\epsilon > 0$, and let

$$a_1, \ldots, a_s, b_1, \ldots, b_t \in \Delta(q)$$

such that $\overline{\Delta(\frac{1}{2})}$, $\overline{\Delta(a_i,\epsilon)}$, $\overline{\Delta(b_j,\epsilon)}$ $(1 \leq i \leq s,\ 1 \leq j \leq t)$ are all mutually disjoint, where $\overline{\Delta(a_i,\epsilon)}$ denotes the topological closure of the disc $\Delta(a_i,\epsilon)$ with center a_i and radius ϵ . We denote by $N(A,\epsilon,\{a_i\},\{b_j\})$ the set of all

$z \in \Delta$ such that $\|f_z\|_{V_1} \leq A$ and $\left\|\frac{1}{f_z}\right\|_{V_2} \leq A$, where

$$V_1 = \Delta(q) - \bigcup_{j=1}^{t} \overline{\Delta(b_j,\epsilon)}$$

and

$$V_2 = \Delta(q) - \bigcup_{i=1}^{s} \overline{\Delta(a_i,\epsilon)}$$

We say that the collection $(A,\epsilon,\{a_i\},\{b_j\})$ is rational if A and ϵ are rational numbers and each of a_i, b_j $(1 \leq i \leq s,\ 1 \leq j \leq t)$ has both rational real and imaginary parts. Let W be a nonempty open subset of Δ . Since each $N(A,\epsilon,\{a_i\},\{b_j\})$ is a closed subset of Δ and each point of Δ belongs to some $N(A,\epsilon,\{a_i\},\{b_j\})$ with rational $(A,\epsilon,\{a_i\},\{b_j\})$, by Baire category theorem, we conclude that there exist a nonempty open subset W' of W and a rational collection $(A,\epsilon,\{a_i\},\{b_j\})$ such that

$$W' \subset N(A,\epsilon,\{a_i\},\{b_j\}) .$$

By (1.11), $W' \subset G$. Hence G is dense.

Let $B = \Delta - G$. We have to show that $B = \emptyset$. Suppose $B \neq \emptyset$. By applying Baire category theorem to B , we obtain an open subset W of Δ such that $W \cap B \neq \emptyset$ and

$$W \cap B \subset N(A,\epsilon,\{a_i\},\{b_j\})$$

for some rational collection $(A,\epsilon,\{a_i\},\{b_j\})$. Since $W \not\subset G$, it follows that there exists $z_0 \in W \cap B$ such that

$r(z_0) < q$.

Choose $\eta > 0$ such that $\Delta(z_0,\eta) \subset\subset W$. Let $\{G_\alpha\}$ be the set of all topological components of G . We claim that there exists $0 < \delta < \frac{\eta}{2}$ such that the diameter of G_α is $< \frac{\eta}{2}$ whenever G_α intersects $\Delta(z_0,\delta)$. For, otherwise there exists, for every $1 \leq \nu < \infty$, some α_ν such that the diameter of G_{α_ν} is $\geq \frac{\eta}{2}$ and G_{α_ν} intersects $\Delta(z_0,\frac{1}{\nu})$. We have $z_0 \in \partial\left(\bigcup_\nu G_{\alpha_\nu}\right)$. Since $r(z) \geq q$ on G , $u(z) \leq -\log q$ on G . By (1.9), $u(z_0) \leq -\log q$, which contradicts $r(z_0) < q$. The claim is proved.

Let z_1 be an arbitrary point of $\Delta(z_0,\delta) - B$. z_1 belongs to some G_α . Since $z_1 \in G_\alpha \cap \Delta(z_0,\delta)$, the diameter of G_α is $< \frac{\eta}{2}$. Since $\delta < \frac{\eta}{2}$, $G_\alpha \subset \Delta(z_0,\delta) \subset\subset W$. It follows that

$$\partial G_\alpha \subset W \cap B \subset N(A,\epsilon,\{a_i\},\{b_j\}) .$$

By (1.11)(i),

$$G_\alpha \subset N(A,\epsilon,\{a_i\},\{b_j\}) .$$

Since z_1 is an arbitrary point of $\Delta(z_0,\delta) - B$,

$$\Delta(z_0,\delta) - B \subset N(A,\epsilon,\{a_i\},\{b_j\}) .$$

Since $N(A,\epsilon,\{a_i\},\{b_j\})$ is closed and B is nowhere dense in Δ ,

$$\Delta(z_0,\delta) \subset N(A,\epsilon,\{a_i\},\{b_j\}) .$$

By (1.11), f is meromorphic on $\Delta(z_0,\delta) \times \Delta(q)$ which contradicts $z_0 \notin G$. We conclude that $B = \emptyset$. Q.E.D.

(1.13) **Theorem** *(on Separate Meromorphicity). Suppose* Z *is a subset of* $\Delta \times \Delta$ *such that* $(\{z\} \times \Delta) \cap Z$ *and* $(\Delta \times \{w\}) \cap Z$ *are both locally finite in* $\Delta \times \Delta$ *for all* $z, w \in \Delta$. *Suppose* f *is a function on* $\Delta \times \Delta - Z$ *such that* $f(z,\cdot)$ *and* $f(\cdot,w)$ *are respectively meromorphic on* $\{z\} \times \Delta$ *and* $\Delta \times \{w\}$ *for all* $z, w \in \Delta$. *Then* f *is the restriction of a meromorphic function on* $\Delta \times \Delta$.

Proof. Fix arbitrarily $0 < q < 1$. We need only show that f is meromorphic on $\Delta \times \Delta(q)$. We use the notation $N(A,\epsilon,\{a_i\},\{b_j\})$ in the same sense as in (1.12). Since $f(\cdot,w)$ is meromorphic on $\Delta \times \{w\}$ for all $w \in \Delta$, $N(A,\epsilon,\{a_i\},\{b_j\})$ is a closed subset of Δ. By Baire category theorem, there exists a nonempty open subset U of such that $U \subset N(A,\epsilon,\{a_i\},\{b_j\})$ for some A, ϵ, $\{a_i\}$, $\{b_j\}$. By using iterated Cauchy integrals, we conclude that f is meromorphic on $U \times \Delta(q)$. By (1.12), f is meromorphic on $\Delta \times \Delta(q)$. Q.E.D.

(1.14) **Remark.** We have formulated (1.12) and (1.13) only in the case of dimension 2. They are also true in the case of any dimension ≥ 2. The proof in the higher-dimensional case uses induction on n. In going from dimension n to dimension $n+1$, we have to define $r(z)$ as the sup of all ρ such that f can be extended meromorphically to some open neighborhood of $\{z\} \times (\Delta(\rho))^n$ in $\mathbb{C}^{n+1}$. Instead of $N(A,\epsilon,\{a_i\},\{b_j\})$, we have to introduce $N(A,V_1,V_2)$ which denotes the set of all $z \in \Delta$ such that $\|f_z\|_{V_1}$ and $\left\|\frac{1}{f_z}\right\|_{V_2} \leqq A$, where V_1, V_2 are open subsets of $\mathbb{C}^n$ satisfying the following property: there exist affine transformations $S_{i_\nu}^{(\nu)}$, T_j of $\mathbb{C}^n$ $(\nu = 1,2;\ 1 \leqq i_\nu \leqq k_\nu;\ 1 \leqq j \leqq \ell)$ whose coefficients have rational real and imaginary parts such that

(i) $V_\nu = \bigcup_{i_\nu=1}^{k_\nu} S_{i_\nu}^{(\nu)}(\Delta^n)$ $(\nu = 1,2)$,

(ii) $T_j(\Delta^{n-p_j} \times \partial(\Delta^{p_j})) \subset V_1 \cup V_2$ for some $p_j \geqq 2$ $(1 \leqq j \leqq \ell)$,

(iii) $\Delta(q)^n \subset \bigcup_{j=1}^{\ell} T_j(\Delta^n)$, where q is a prescribed number in $(1,r)$.

Every $z \in \Delta$ belongs to one such $N(A,V_1,V_2)$ with rational A. The reason why such a complicated definition of $N(A,V_1,V_2)$ is needed is that the pole-set and the zero-set of a meromorphic function on a polydisc of dimension > 1 may intersect. In general, we cannot choose a relatively compact open subset of the complement of the pole-set and a relatively compact open subset of the complement of the zero-set so that the union of these two open subsets equals the whole polydisc. However, we can choose the two open subsets so that the envelope of holomorphy of their union contains a prescribed slightly smaller polydisc, because the intersection of the pole-set and the zero-set is of codimension ≥ 2. This is precisely the idea behind the definition of $N(A,V_1,V_2)$. If U is an open subset of Δ such that $U \subset N(A,V_1,V_2)$, then f is meromorphic on $U \times \Delta(q)$.

(1.15) Remark. The arguments used in the proof of (1.12) and (1.13) and their higher-dimensional cases work also for holomorphic functions. In the case of holomorphic functions, the arguments can be greatly simplified. Of course, these results on holomorphic functions are more readily obtained by the usual proof using power series expansions and the lemma that, if u_k is a sequence of subharmonic functions on Ω uniformly bounded from above satisfying $\overline{\lim}_{k \to \infty} u_k \leq A$ for $z \in \Omega$, then $\overline{\lim}_{k \to \infty} \sup_{z \in K} u_k(z) \leq A$ for every compact subset K of Ω.

Appendix of Chapter 1

The $\frac{1}{4}$-Theorem of Koebe-Bieberbach

(1.A.1) Lemma. *Suppose* $f(z) = \sum_{\nu=-\infty}^{\infty} a_\nu z^\nu$ *is a holomorphic function on a domain in* $\mathbb{C}$ *which contains* $\{|z| = r\}$. *Suppose* $f(z)$ *describes a simple closed curve* Γ *once in the clockwise direction as* z *describes* $\{|z| = r\}$ *once in the counter-clockwise direction. Then the area enclosed by* Γ *is*

$$-\pi \sum_{\nu=-\infty}^{\infty} \nu |a_\nu|^2 r^{2\nu} .$$

Proof. Let u and v be respectively the real and imaginary parts of f. The area enclosed by Γ is

$$-\frac{1}{2}\int_{|z|=r} udv - vdu = -\frac{i}{2}\int_{|z|=r} f\, d\bar{f}$$

$$= -\frac{i}{2}\int_{|z|=r} f\, \overline{f'}\, d\bar{z}$$

$$= -\frac{i}{2}\int_{|z|=r} \Big(\sum_{\mu=-\infty}^{\infty} a_\mu z^\mu\Big)\Big(\sum_{\nu=-\infty}^{\infty} \nu \bar{a}_\nu \frac{r^{2\nu-2}}{z^{\nu-1}}\Big)\Big(-\frac{r^2 dz}{z^2}\Big)$$

$$= -\pi \sum_{\nu=-\infty}^{\infty} \nu |a_\nu|^2 r^{2\nu} . \qquad \text{Q.E.D.}$$

(1.A.2) Lemma. *Suppose* $f(z) = z + \sum_{\nu=2}^{\infty} a_\nu z^\nu$ *is a univalent holomorphic function on* Δ. *Then* $|a_2| \leq 2$.

Proof. The function $F(z) := f(z^2)^{1/2}$ with the expansion

$$F(z) = z + \frac{1}{2} a_2 z^3 + \ldots$$

is a well-defined univalent holomorphic function on Δ. (This can be seen by composing $w = f(z^2)$ with the multi-valued holomorphic function $\zeta = \sqrt{w}$ and by observing that, as z goes around 0 in a small circle, the two images of the multi-valued composite function trace out two closed paths.)

$$\frac{1}{F(z)} = \frac{1}{z} - \frac{1}{2} a_2 z + \sum_{\nu=2}^{\infty} b_\nu z^\nu$$

Since, for $0 \leqq r < 1$, the area enclosed by the image of $\{|z| = r\}$ by $\frac{1}{F(z)}$ is nonnegative, by (1.A.1) we have

$$\frac{1}{r^2} - \left(\frac{1}{4}|a_2|^2 r^2 + \sum_{\nu=2}^{\infty} \nu |b_\nu|^2 r^{2\nu} \right) \geqq 0 .$$

Hence $|a_2| \leqq 2$. Q.E.D.

(1.A.3) Theorem (The $\frac{1}{4}$-Theorem of Koebe-Bieberbach). *Suppose* f *is a univalent holomorphic function on* Δ *such that* $f(0) = 0$ *and* $f'(0) = 1$. *Then* $f(\Delta) \supset \Delta(\frac{1}{4})$

Proof. Suppose the contrary. Then there exists $b \in \Delta(\frac{1}{4}) - f(\Delta)$. Let

$$f(z) = z + \sum_{\nu=2}^{\infty} a_\nu z^\nu .$$

The function

$$\frac{bf(z)}{b-f(z)} = z + (a_2 + \frac{1}{b}) z^2 + \ldots$$

is univalent and holomorphic on Δ . By (1.A.2), $|a_2| \leqq 2$ and $\left|a_2 + \frac{1}{b}\right| \leqq 2$. Hence $\left|\frac{1}{b}\right| \leqq 2 + |a_2| \leqq 4$ and $|b| \geqq \frac{1}{4}$, contradicting $b \in \Delta(\frac{1}{4})$. Q.E.D.

CHAPTER 2

EXTENSION OF SUBVARIETIES AND HOLOMORPHIC MAPS

In this chapter we will prove the following results: (I) the theorem of Thullen-Remmert-Stein, (II) Bishop's theorem, (III) Rothstein's theorem, (IV) the extension of subvarieties across $\mathbb{R}^n$, and (V) the application of Bishop's theorem to the extension of holomorphic maps.

(I). Theorem of Thullen-Remmert-Stein

The theorem of Thullen-Remmert-Stein depends on a theorem of Radó [13]. We give here a simple elegant proof of Rado's theorem due to R. Kaufman [10] which uses only the maximum modulus principle.

(2.1) Theorem (Radó). *Let f be a continuous function on an open subset G of $\mathbb{C}^n$. If f is holomorphic on $\{z \in G \mid f(z) \neq 0\}$, then f is holomorphic on G.*

Proof. By the iterated Cauchy's integral formula, a continuous function which is separately holomorphic is holomorphic. Hence we can assume without loss of generality that G is the open unit 1-disc Δ and that f is continuous on $\bar{\Delta}$. Let $U = \{z \in \Delta \mid f(z) \neq 0\}$. First we show that

(#) if g is a continuous function on $\bar{U}$ and is holomorphic on U, then $|g(a)| \leq \sup_{z \in (\partial\Delta) \cap (\partial U)} |g(z)|$ for all $a \in U$.

This follows from taking the nth root on both sides of the inequality

$$|g^n(a)f(a)| \leq \sup_{z \in \partial U} |g^n(z)f(z)| = \sup_{z \in (\partial\Delta) \cap (\partial U)} |g^n(z)f(z)|$$

(which comes from the maximum modulus principle and the fact that $f = 0$ on $\Delta \cap \partial U$) and then letting $n \longrightarrow \infty$.

U is dense in Δ, otherwise there exists a sequence $\{a_k\}$ in $\Delta - \overline{U}$ approaching a point of $\Delta \cap \partial U$ as limit and the sequence $\frac{1}{z-a_k}$ converges uniformly on $(\partial\Delta) \cap (\partial U)$ but is not uniformly bounded on U, contradicting (#).

Choose polynomials p_k such that $|\mathrm{Re}(p_k - f)| \leq \frac{1}{k}$ on $\partial\Delta$. Then $\left|e^{p_k - f}\right| \leq e^{1/k}$ on $\partial\Delta$ and $\left|e^{f-p_k}\right| \leq e^{1/k}$ on $\partial\Delta$. By (#), $\left|e^{p_k - f}\right| \leq e^{1/k}$ on U and $\left|e^{f-p_k}\right| \leq e^{1/k}$ on U. Since U is dense in Δ, $|\mathrm{Re}(p_k - f)| \leq \frac{1}{k}$ on Δ. Hence $\mathrm{Re}\, f$ is harmonic on Δ. Likewise $\mathrm{Im}\, f$ is harmonic on Δ. f is real-analytic on Δ. Since $\frac{\partial f}{\partial \bar{z}} = 0$ on U and U is dense in Δ, $\frac{\partial f}{\partial \bar{z}} = 0$ on Δ. Q.E.D.

(2.2) Lemma. *Suppose A is a subvariety of dimension $\leq k$ in an open subset Ω of $\mathbb{C}^n$ and $1 \leq \ell \leq k$. Let $G_{n-\ell}(\mathbb{C}^n)$ be the Grassmannian of all $(n-\ell)$-dimensional planes in $\mathbb{C}^n$ passing through 0. Let R be the set of all $T \in G_{n-\ell}(\mathbb{C}^n)$ such that $\dim A \cap (x+T) \geq k-\ell+1$ for some $x \in A$ (where $x+T$ is the translate of T by x). Then R is thin in $G_{n-\ell}(\mathbb{C}^n)$.*

Proof. (a) Let B be the set of all

$$(x,T) \in A \times G_{n-\ell}(\mathbb{C}^n)$$

such that

$$\dim_x A \cap (x+T) \geq k - \ell + 1 .$$

We are going to show that B is a subvariety of $A \times G_{n-\ell}(\mathbb{C}^n)$. For

$$1 \leq i_1 < \dots < i_{n-\ell} \leq n$$

let $U_{i_1 \dots i_{n-\ell}}$ be the set of all $T \in G_{n-\ell}(\mathbb{C}^n)$ such that

$$\dim T \cap \{ z_{i_1} = \dots = z_{i_{n-\ell}} = 0 \} = 0 .$$

$$\{ U_{i_1 \dots i_{n-\ell}} \}_{1 \leq i_1 < \dots < i_{n-\ell} \leq n}$$

is an open covering of $G_{n-\ell}(\mathbb{C}^n)$. We need only show that $B \cap (A \times U_{i_1 \dots i_{n-\ell}})$ is a subvariety of $A \times U_{i_1 \dots i_{n-\ell}}$. We can assume without loss of generality that $i_\nu = \ell+\nu$ for $1 \leq \nu \leq n-\ell$. Define

$$\alpha: A \times U_{\ell+1,\dots,n} \longrightarrow \mathbb{C}^\ell$$

as follows:

$$\alpha(x,T) = (z_1, \dots, z_\ell)$$

such that

$$(z_1, \dots, z_\ell, \overbrace{0, \dots, 0}^{n-\ell}) \in x + T .$$

α is well-defined and holomorphic, because, if T is given by

$$\sum_{j=1}^{n} \lambda_{ij} z_j = 0 \quad (1 \leqq i \leqq \ell)$$

and

$$x = (z_1^0, \ldots, z_n^0) ,$$

then $\alpha(x,T)$ equals the product of

$$\left(\sum_{j=1}^{n} \lambda_{1j} z_j^0, \ \ldots, \ \sum_{j=1}^{n} \lambda_{\ell j} z_j^0 \right)$$

and the inverse transpose of $(\lambda_{ij})_{1 \leqq i,j \leqq \ell}$. Let

$$\tilde{\alpha}: A \times U_{\ell+1,\ldots,n} \longrightarrow \mathbb{C}^{\ell} \times U_{\ell+1,\ldots,n}$$

be defined by $\tilde{\alpha}(x,T) = (\alpha(x),T)$. Since $B \cap (A \times U_{\ell+1,\ldots,n})$ equals the set of all $a \in A \times U_{\ell+1,\ldots,n}$ such that

$$\dim_a \tilde{\alpha}^{-1}\tilde{\alpha}(a) \geqq k - \ell + 1 ,$$

it follows that $B \cap (A \times U_{\ell+1,\ldots,n})$ is a subvariety of $A \times U_{\ell+1,\ldots,n}$ (see the Appendix of Chapter 2 for this and also for statements concerning the properties of rank in this proof). Hence B is a subvariety of $A \times G_{n-\ell}(\mathbb{C}^n)$.

(b) Let

$$\sigma: A \times G_{n-\ell}(\mathbb{C}^n) \longrightarrow A$$

and

$$\tau: A \times G_{n-\ell}(\mathbb{C}^n) \longrightarrow G_{n-\ell}(\mathbb{C}^n)$$

be the natural projections. We have $R = \tau(B)$. Since the holomorphic image of a complex space is always a countable union of subvarieties of open subsets, to finish the proof we need only show that R has no interior in $G_{n-\ell}(\mathbb{C}^n)$.

(c) We are going to reduce the problem to the special situation of $k = \ell$. Take $T \in G_{n-\ell}(\mathbb{C}^n)$ and an open neighborhood U of T in $G_{n-\ell}(\mathbb{C}^n)$. Fix $T' \in G_{n-k}(\mathbb{C}^n)$ such that $T' \subset T$. There exists $T'' \in G_{k-\ell}(\mathbb{C}^n)$ such that $T'' \subset T$ and $T' \cap T'' = \{0\}$. There exists an open neighborhood U' of T' in $G_{n-k}(\mathbb{C}^k)$ such that, for $S' \in U'$, $S' \cap T'' = \{0\}$ and the unique element S of $G_{n-\ell}(\mathbb{C}^n)$ which contains S' and T'' belongs to U . Suppose we have shown that, when $k = \ell$, R has no interior in $G_{n-k}(\mathbb{C}^n)$. Then there exists $S' \in U'$ such that

$$\dim A \cap (x+S') \leq 0$$

for every $x \in A$. Let S be the unique element of $G_{n-\ell}(\mathbb{C}^n)$ which contains S' and T'' . Then $S \in U$ and

$$\dim A \cap (x+S) \leq k - \ell$$

for every $x \in A$. In the rest of the proof we assume $k = \ell$ and without loss of generality assume that Ω is convex.

(d) Consider first the special case $k = n-1$. $G_1(\mathbb{C}^n) = \mathbb{P}_{n-1}$. Let $[w_1, \ldots, w_n]$ be the homogeneous coordinates of $\mathbb{P}_{n-1}$. We have to show that $\operatorname{rank}(\tau | B) \leq n-2$. Suppose the contrary. Then there exist an open neighborhood E of some point $y_0 \in \mathbb{P}_{n-1}$ and a holomorphic map

$\varphi\colon E \longrightarrow B$ such that $\tau\varphi$ = the identity map of E. Without loss of generality we can assume that

$$y_0 = [\overbrace{0, \ldots, 0}^{n-1},1] .$$

Let U be the set of all

$$[w_1, \ldots, w_n] \in \mathbb{P}_{n-1}$$

with $w_n \neq 0$. Define

$$\psi\colon (U \cap E) \times \mathbb{C} \longrightarrow \mathbb{C}^n$$

by

$$\psi([w_1, \ldots, w_n],\xi) = \sigma\varphi([w_1, \ldots, w_n]) + \left(\xi\frac{w_1}{w_n}, \ldots, \xi\frac{w_{n-1}}{w_n},\xi\right) .$$

ψ has rank n, because the map $\tilde{\psi}\colon (U \cap E) \times \mathbb{P} \longrightarrow \mathbb{P}_n$ defined by

$$\tilde{\psi}([w_1, \ldots, w_n],[\xi_0,\xi_1])$$

$$= \left[\ \xi_1\sigma\varphi([w_1, \ldots, w_n]) + \left(\xi_0\ \frac{w_1}{w_n}, \ \ldots, \ \xi_0\frac{w_{n-1}}{w_n},\xi_0\right),\xi_1\right]$$

extends ψ and

$$\tilde{\psi}^{-1}\tilde{\psi}([w_1, \ldots, w_n],[1,0]) = ([w_1, \ldots, w_n],[1,0]) .$$

There exist a nonempty open subset W of $U \cap E$ and a nonempty open subset D of $\mathbb{C}$ such that ψ maps $W \times D$ biholomorphically onto an open subset of Ω. Since Ω is convex,

$(x+T) \cap \Omega \subset A$ for every $(x,T) \in B$. It follows that $\Psi(W \times D) \subset A$, contradicting $\dim A \leqq n-1$.

(e) For the general case where k may not be $n-1$, we use induction on $n-k$. Fix $T \in R$ and an open neighborhood H of T in $G_{n-k}(\mathbb{C}^n)$. Let T be defined by

$$\sum_{j=1}^{n} \lambda_{ij} z_j = 0 \quad (1 \leqq i \leqq k) .$$

Without loss of generality we can assume that $(\lambda_{ij})_{1 \leqq i,j \leqq k}$ is nonsingular. Take arbitrarily $\epsilon > 0$. By applying the special case of $k = n-1$ to the 1-dimensional plane $T \cap \{z_{k+1} = \dots = z_{n-1} = 0\}$, we obtain $(\mu_{ij})_{\substack{1 \leqq i \leqq n-1 \\ 1 \leqq j \leqq n}}$ such that

(i) $|\mu_{ij} - \lambda_{ij}| < \epsilon$ for $1 \leqq i \leqq k$, $1 \leqq j \leqq n$,

(ii) $|\mu_{ij} - \delta_{ij}| < \epsilon$ for $k < i < n$, $1 \leqq j \leqq n$ (where δ_{ij} is the Kronecker delta),

(iii) the fibers of the map $\Phi : A \longrightarrow \mathbb{C}^{n-1}$ defined by

$$(z_1, \dots, z_n) \longrightarrow \left(\sum_{j=1}^{n} \mu_{ij} z_j, \dots, \sum_{j=1}^{n} \mu_{n-1,j} z_j\right)$$

have dimensions $\leqq 0$.

$\Phi(A) = \bigcup_{\gamma=1}^{\infty} A_\gamma$, where A_γ is a subvariety of dimension $\leqq k$ in a convex open subset Ω_γ of $\mathbb{C}^{n-1}$.

By induction hypothesis there exists $(\nu_{\ell i})_{\substack{1 \leqq \ell \leqq k \\ 1 \leqq i \leqq n-1}}$ such that

(i) $|\nu_{\ell i} - \delta_{\ell i}| < \epsilon$ for $1 \leqq \ell \leqq k$ and $1 \leqq i \leqq n-1$,

(ii) for every γ the fibers of the map $A_\gamma \longrightarrow \mathbb{C}^k$ defined by

$$(\zeta_1, \ldots, \zeta_{n-1}) \longrightarrow \left(\sum_{j=1}^{n-1} \nu_{1j}\zeta_j, \ldots, \sum_{j=1}^{n-1} \nu_{n-1,j}\zeta_j\right)$$

have dimensions $\leqq 0$.

Let $\lambda'_{\ell j} = \sum_{i=1}^{n-1} \nu_{\ell i}\mu_{ij}$ and let $T' \in G_{n-k}(\mathbb{C}^n)$ be defined by $\sum_{j=1}^{n} \lambda'_{\ell j} z_j = 0$ $(1 \leqq i \leqq k)$. Then $T' \notin R$ and, for ϵ sufficiently small, $T \in H$. Q.E.D.

<u>Remark</u>. For the proof of the theorem of Thullen-Remmert-Stein, the following weaker version of (2.2) suffices: For every fixed $x_0 \in \mathbb{C}^n$, the set of all $T \in G_{n-k}(\mathbb{C}^n)$ such that $\dim A \cap (x_0+T) \geqq 1$ is of first category in $G_{n-k}(\mathbb{C}^n)$. This weaker version can be proved much more easily. However, the full strength of (2.2) is needed later in the proof of Rothstein's theorem.

(2.3) <u>Lemma</u>. <u>Suppose</u> Ω <u>is a Stein open subset of</u> $\mathbb{C}^n$ <u>and</u> $\varphi: \Omega \longrightarrow \mathbb{C}^k$ <u>is a holomorphic map. Suppose</u> U <u>is a connected open subset of</u> $\mathbb{C}^k$, Ω' <u>is an open subset of</u> Ω , <u>and</u> X <u>is</u>

a subvariety in Ω' such that $\varphi|X$ makes X an analytic cover over U. Suppose $\tilde{U}$ is a connected open subset of $\mathbb{C}^k$ containing U.

(A) If X_i^* is a subvariety in an open neighborhood Ω_i^* of Ω' in Ω such that $X_i^* \cap \Omega' = X$ and $\varphi|X_i^*$ makes X_i^* an analytic cover over $\tilde{U}$ $(i = 1,2)$, then $X_1^* = X_2^*$.

(B) Suppose one of the following three conditions (a), (b), (c) is satisfied.

(a) (i) There exists $K \subset \Omega$ such that $K \cap \varphi^{-1}(\tilde{U}) \longrightarrow \tilde{U}$ induced by φ is proper.

(ii) $X \subset K \cap \varphi^{-1}(U)$.

(iii) Every holomorphic function on U which is locally bounded on $\tilde{U}$ can be extended to a holomorphic function on $\tilde{U}$.

(b) Every holomorphic function on U can be extended to a holomorphic function on $\tilde{U}$.

(c) (i) There exist a connected open subset D of $\mathbb{C}^{k-1}$ and $0 < \alpha < \beta$ such that $U = D \times \left(\Delta(\beta) - \overline{\Delta(\alpha)}\right)$ and $\tilde{U} = D \times \Delta(\beta)$

(ii) There exists a thick set A in D satisfying the following property: for $t \in A$, $X_t := X \cap \varphi^{-1}(\{t\} \times \mathbb{C})$ can be extended to a subvariety $\tilde{X}_t$ in some open neighborhood Ω_t of Ω' in Ω such that $\varphi|\tilde{X}_t$ makes $\tilde{X}_t$ an analytic cover over $\{t\} \times \Delta(\beta)$.

Then X *can be extended to a subvariety* $\tilde{X}$ *in* $\Omega \cap \varphi^{-1}(\tilde{U})$ *such that* $\varphi|\tilde{X}$ *makes* $\tilde{X}$ *an analytic cover over* $\tilde{U}$.

Proof. Let $\psi: \Omega \longrightarrow \mathbb{C}^N$ be a proper holomorphic embedding of Ω . The diagram

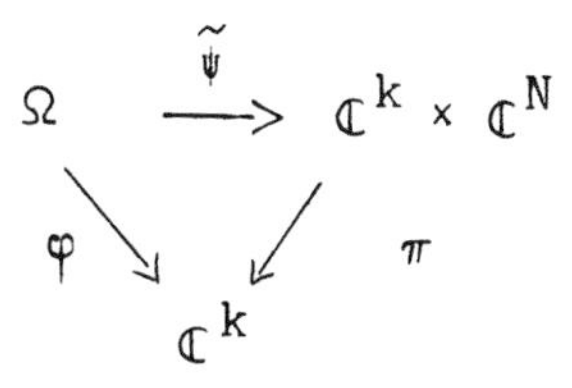

is commutative, where π is the projection onto the first factor and $\tilde{\psi}(x) = (\varphi(x), \psi(x))$. By replacing Ω by $\mathbb{C}^k \times \mathbb{C}^N$ and X by $\tilde{\psi}(X)$, we can assume without loss of generality that $\Omega = \mathbb{C}^k \times \mathbb{C}^N$ and $\varphi = \pi$. Let λ be the number of sheets of the analytic cover X over U .

The following two properties of analytic covers are well-known and can easily be proved respectively by using elementary symmetric functions and the Cramer's rule:

(i) If f is a holomorphic function on X , then there exists uniquely

$$P_f(z;Z) = Z^\lambda + \sum_{i=0}^{\lambda-1} a_i(z)Z^i$$

(where a_i is a holomorphic function on U , $0 \leq i < \lambda$) such that $P_f(\pi(x);f(x)) = 0$ for every $x \in X$. The functions a_i are uniformly bounded on an open subset W of U if f is uniformly bounded on $X \cap \pi^{-1}(W)$. If L is a positive-dimensional plane in $\mathbb{C}^k$ such that no nonempty open subset of L is contained in the minimal critical set of the analytic cover $X \longrightarrow U$, then the coefficients of the polynomial $P_{f|X \cap \pi^{-1}(L \cap U)}(z;Z)$ in Z associated to the function $f|X \cap \pi^{-1}(L \cap U)$ on the analytic cover $X \cap \pi^{-1}(L \cap U) \longrightarrow L \cap U$ are precisely the restrictions to $L \cap U$ of the coefficients of the polynomial $P_f(z;Z)$ in Z .

(ii) If f and g are holomorphic functions on X and f assumes λ distinct values on $X \cap \pi^{-1}(z')$ for some $z' \in U$, then there exists uniquely

$$T_{f,g}(z;Z) = \sum_{i=0}^{\lambda-1} b_i(z)Z^i$$

(where $b_i(z)$ is a holomorphic function on U) such that

$$g(x)P_f'(\pi(x);f(x)) = T_{f,g}(\pi(x);f(x))$$

for every $x \in X$ (where $P_f'(z;Z)$ is the derivative of $P_f(z;Z)$ with respect to Z) . The functions b_i are uniformly bounded on an open subset W of U if both f and g are uniformly bounded on $X \cap \pi^{-1}(W)$. If L is a positive-dimensional plane in $\mathbb{C}^k$ such that in every topological component of $L \cap U$ there exists a point z^* with f assuming λ distinct values on $X \cap \pi^{-1}(z^*)$, then the coefficients of the polynomial $T_{f|X \cap \pi^{-1}(L \cap U), g|X \cap \pi^{-1}(L \cap U)}(z;Z)$ in Z associated to the holomorphic functions $f|X \cap \pi^{-1}(L \cap U)$ and $g|X \cap \pi^{-1}(L \cap U)$ on the analytic cover $X \cap \pi^{-1}(L \cap U) \longrightarrow L \cap U$ are precisely the restrictions to $L \cap U$ of the coefficients of the polynomial $T_{f,g}(z;Z)$ in Z.

We will prove (B) first. Let $w_1, \ldots, w_N$ be the coordinates of $\mathbb{C}^N$. Choose $\mu_1, \ldots, \mu_N \in \mathbb{C}$ such that $f := \mu_1 w_1 + \ldots + \mu_N w_N$ assumes λ distinct values on $X \cap \pi^{-1}(z')$ for some $z' \in U$. Let C be the set of all the coefficients of the polynomials $P_{f|X}(z;Z)$, $P_{w_j|X}(z;Z)$, and $T_{f|X, w_j|X}(z;Z)$ $(1 \leq j \leq N)$ in Z. We are going to show that every element of C can be extended to a holomorphic function on $\tilde{U}$. If (a) is satisfied, since $X \subset K \cap \varphi^{-1}(U)$ and $K \cap \varphi^{-1}(\tilde{U}) \longrightarrow \tilde{U}$ is proper, every element of C is locally bounded on $\tilde{U}$ and therefore can be extended to a holomorphic function on $\tilde{U}$. When (b) is satisfied, clearly every element of C can be extended to a holomorphic function on $\tilde{U}$. Suppose (c) is satisfied. Let Y be the set of all $x \in U$ such that f assumes less than λ distinct values on $X \cap \pi^{-1}(x)$. Y is a subvariety of pure codimension 1 in U. Let $\sigma: Y \longrightarrow D$ be induced by the natural projection $D \times (\Delta(\beta) - \overline{\Delta(\alpha)}) \longrightarrow D$. Let Y' be the closure of the set of all $x \in Y$ such that $\operatorname{rank}_x \sigma < k-1$. Then $\sigma(Y')$ is thin in D (see the Appendix of Chapter 2 for the definition and properties of rank). For $t \in D - \sigma(Y')$ let C_t be the set of all the coefficients of the polynomials $P_{f|X_t}(z;Z)$, $P_{w_j|X_t}(z;Z)$, $T_{f|X_t, w_j|X_t}(z;Z)$ in Z associated to the holomorphic functions $f|X_t$, $w_j|X_t$ on the analytic cover

$$\pi|X_t: X_t \longrightarrow \{t\} \times (\Delta(\beta) - \overline{\Delta(\alpha)})$$

$(1 \leq j \leq N)$. Elements of C_t are precisely the restrictions of elements of C to $\{t\} \times (\Delta(\beta) - \overline{\Delta(\alpha)})$ for $t \in D-\sigma(Y')$. Since for $t \in A$ the analytic cover $X_t \longrightarrow \{t\} \times (\Delta(\beta) - \overline{\Delta(\alpha)})$ can be extended to the analytic cover $\tilde{X}_t \longrightarrow \{t\} \times \Delta(\beta)$ every element of C_t can be extended to a holomorphic function on $\{t\} \times \Delta(\beta)$ for $t \in A-\sigma(Y')$. By (1.2) every element of C can be extended to a holomorphic function on $\tilde{U}$. By using the extensions of elements of C , we obtain new polynomials $\tilde{P}_{f|X}(z;Z)$, $\tilde{P}_{w_j|X}(z;Z)$, $\tilde{T}_{f|X,w_j|X}(z;Z)$ whose coefficients in Z are holomorphic functions on $\tilde{U}$. The set $\tilde{X}$ of all $(z;w_1, \ldots, w_N) \in \tilde{U} \times \mathbb{C}^N$ satisfying

$$(\dagger) \quad \begin{cases} \tilde{P}_{f|X}(z;f) = 0 & \\ \tilde{P}_{w_j|X}(z;w_j) = 0 & (1 \leq j \leq N) \\ w_j\tilde{P}'_{f|X}(z;f) = T_{f|X,w_j|X}(z;w_j) & \end{cases}$$

(where $\tilde{P}'_{f|X}(z;Z)$ is the derivative of $\tilde{P}_{f|X}(z;Z)$ with respect to Z) is a subvariety of $\tilde{U} \times \mathbb{C}^N$ extending X such that $\pi|\tilde{X}$ makes $\tilde{X}$ an analytic cover over $\tilde{U}$.

Now we come to the proof of (A). We use the same notations as for the proof of (B). Since $\pi|X_i^*$ makes X_i^* an analytic cover over $\tilde{U}$, every element of C extends to a holomorphic function on $\tilde{U}$. Let E be a proper subvariety of $\tilde{U}$ containing the minimum critical sets of X_1^*, X_2^*, and $\tilde{X}$. Since $X_i^* \subset \tilde{X}$ and for every $z \in \tilde{U}-E$ both $X_i^* \cap \pi^{-1}(z)$ and $\tilde{X} \cap \pi^{-1}(z)$ have λ points, it follows that $X_i^* - \pi^{-1}(E) = \tilde{X} - \pi^{-1}(E)$ $(i = 1,2)$. Since $X_i^* \cap \pi^{-1}(E)$ is nowhere dense in X_i^* and $\tilde{X} \cap \pi^{-1}(E)$ is nowhere dense in $\tilde{X}$, we have $X_i^* = \tilde{X}$ $(i = 1,2)$. Q.E.D.

(2.4) Lemma. *Suppose* G *is an open subset of* $\mathbb{C}^n$, D *is a connected open subset of* $\mathbb{C}^k$, *and* f_i *is a holomorphic function on* $D \times G$ $(i \in I)$. *Let* E *be the set of all* $x \in D \times G$ *such that* $f_i(x) = 0$ *for* $i \in I$. *Suppose* K *is a compact subset of* G *and* D' *is a nonempty open subset of* D . *If* A *is a subvariety of* $D \times G - E$ *such that* $A \subset D \times K$ *and* $A \cap (D' \times G) = \emptyset$, *then* $\dim A < k$.

Proof. Suppose $\dim A \geqq k$. We are going to derive a contradiction. From considering a boundary point of the image under the projection $D \times G \longrightarrow D$ of a branch of A of dimension $\geqq k$, it is clear that we can confine ourselves to the special case $D = \Delta^k$ and $D' = (\Delta(\epsilon))^k$ for some $0 < \epsilon < 1$. By taking 1-dimensional linear subspaces of $\mathbb{C}^k$, we reduce the general case to the special case $k = 1$. We can also assume that A is irreducible.

Take $x_0 \in A$. For some $i \in I$, $f_i(x_0) \neq 0$. Take $\epsilon < \delta < 1$ such that $x_0 \in \Delta(\delta) \times G$. There exists $x^* \in A \cap (\overline{\Delta(\delta)} \times G)$ such that

$$\left|f_i(x^*)\right| = \sup_{x \in A \cap (\overline{\Delta(\delta)} \times G)} |f_i(x)| ,$$

because $\sup_{x \in A \cap (\overline{\Delta(\delta)} \times G)} |f_i(x)|$ is equal to the sup of $|f_i|$ on

$$A \cap (\overline{\Delta(\delta)} \times K) \cap \{|f_i| \geq |f_i(x_0)|\}$$

which is a compact set. If f_i is not constant on A, by the maximum modulus principle for the function $f_i|A$, we have $x^* \in (\partial\Delta(\delta)) \times G$. If f_i is constant on A, A is a subvariety of $D \times G$. Since A is noncompact, $A \cap ((\partial\Delta(\delta)) \times G) \neq \emptyset$ and we can assume (by replacing x^* by another point if necessary) that $x^* \in (\partial\Delta(\delta)) \times G$.

Let z be the coordinate of D. There exists m_0 such that

$$\left|\frac{f_i(x_0)}{z^m(x_0)}\right| > \left|\frac{f_i(x^*)}{z^m(x^*)}\right|$$

for $m \geq m_0$, because $|z(x_0)| < \delta = |z(x^*)|$. We can choose $m \geq m_0$ such that $\frac{f_i}{z^m}$ is not constant on A, otherwise both f_i and z are constant on A and A is a compact subvariety of $D \times G$. There exists $\tilde{x} \in A \cap (\Delta(\delta) \times G)$ such that

$$\left|\frac{f_i(\tilde{x})}{z^m(\tilde{x})}\right| = \sup_{x \in A \cap (\overline{\Delta(\delta)} \times G)} \left|\frac{f_i(x)}{z^m(x)}\right|$$

because $\sup_{x \in A \cap (\overline{\Delta(\delta)} \times G)} \left|\frac{f_i(x)}{z^m(x)}\right|$ is equal to the sup of

$\left|\frac{f_i}{z^m}\right|$ on

$$A \cap ((\overline{\Delta(\delta)} - \Delta(\epsilon)) \times K) \cap \left\{|f_i| \geq \left(\frac{\epsilon}{\delta}\right)^m |f_i(x^*)|\right\}$$

which is compact and because

$$\left|\frac{f_i(x_0)}{z^m(x_0)}\right| > \left|\frac{f_i(x^*)}{z^m(x^*)}\right| \geq \sup_{x \in A \cap ((\partial\Delta(\delta)) \times G)} \left|\frac{f_i(x)}{z^m(x)}\right| .$$

This contradicts the maximum modulus principle for the function $\left.\frac{f_i}{z^m}\right| A$. Q.E.D.

(2.5) <u>Corollary</u>. <u>Suppose</u> G <u>is a Stein open subset of</u> $\mathbb{C}^n$ <u>and</u> E <u>is a proper subvariety of</u> G . <u>Suppose</u> $G'' \subset\subset G'$ <u>are open subsets of</u> G <u>and</u> A <u>is a subvariety of</u> $G'-E$ <u>such that</u> $A \subset G''$. <u>Then</u> $\dim A \leq 0$.

Proof. Identify $\mathbb{C}^n$ with $\{0\} \times \mathbb{C}^n$ in $\mathbb{C}^{n+1}$. The result follows from (2.4) by setting $D = \mathbb{C}$ and $D' = \mathbb{C}-\{0\}$.

Q.E.D.

(2.6) <u>Theorem</u> (Thullen-Remmert-Stein). <u>Suppose</u> G <u>is an open subset of</u> $\mathbb{C}^n$, W <u>is a subvariety of</u> G <u>of dimension</u> $\leq k$, <u>and</u> H <u>is an open subset of</u> G <u>intersecting every branch of</u> W . <u>If</u> V <u>is a subvariety of</u> $(G-W) \cup H$ <u>of pure dimension</u> k , <u>then</u> $G \cap \overline{V}$ <u>is a subvariety of</u> G .

Proof. By using induction on $\dim W$, we reduce the general case to the special case where W is a connected manifold of dimension k . Let $D \neq \emptyset$ be a maximal subdomain of W such that $((G-W) \cup D) \cap \overline{V}$ is a subvariety of $(G-W) \cup D$. We need only prove that D is closed in W . Suppose there exists a point x in the boundary ∂D of D in W . We are going to derive a contradiction. By (2.2) there exists a complex hyperplane P of codimension k passing through x such that $\dim P \cap V \leq 0$ and $\dim P \cap W \leq 0$, we can assume without loss of generality that $G = \Omega \times L$, $W = \Omega \times \{0\}$, and $V \cap (\Omega \times (L-K)) = \emptyset$, where Ω is a connected open subset of $\mathbb{C}^k$, L is an open neighborhood of 0 in $\mathbb{C}^{n-k}$, and K is a compact neighborhood of 0 in L. Let $\pi: \overline{V} \cap (\Omega \times L) \longrightarrow \Omega$ be induced by the natural projection $\Omega \times L \longrightarrow \Omega$. We identify Ω with $\Omega \times \{0\}$. $\pi^{-1}(D)$ is an analytic cover over D.

Choose $\mu_1, \ldots, \mu_{n-k} \in \mathbb{C}$ such that, for some $b \in D$, the function

$$f: = \sum_{i=1}^{n-k} \mu_i w_i$$

is nonzero at every point of $\pi^{-1}(b)$, where $w_1, \ldots, w_{n-k}$ are the last $n-k$ coordinates of $\mathbb{C}^n$. Let $g(z)$ be the holomorphic function on D which is the constant term of the polynomial $P_f(z;Z)$ for the function $f|\pi^{-1}(D)$ on the analytic cover $\pi^{-1}(D) \longrightarrow D$ as defined in the proof of (2.3).

We claim that, if $\{a_\nu\} \subset D$ and $a_\nu \longrightarrow a \in \partial D$ as $\nu \longrightarrow \infty$, then $g(a_\nu) \longrightarrow 0$ as $\nu \longrightarrow \infty$. Suppose the contrary. By replacing $\{a_\nu\}$ by a subsequence, we can assume that $|g(a_\nu)| \geq \epsilon$ for all ν and for some $\epsilon > 0$. Let λ be the number of sheets of the analytic cover $\pi^{-1}(D) \longrightarrow D$. Choose $c \geq 1$ such that $c \geq$ the sup of $|f|$ on K. By (2.5) there exist an open neighborhood U of 0 in L and a connected open neighborhood E of a in Ω such that the sup of $|f|$ on U is $< \frac{\epsilon}{c^{\lambda-1}}$ and $V \cap (E \times \partial U) = \emptyset$. Since $|g(a_\nu)| \geq \epsilon$ for all ν, there exists ν_0 such that $a_\nu \in E$ and $\overline{V} \cap (\{a_\nu\} \times U) = \emptyset$ for $\nu \geq \nu_0$. Let E' be the component of $E \cap D$ containing a_{ν_0}. $\overline{V} \cap (E' \times U) = \emptyset$, otherwise, since $\overline{V} \cap (E' \times U) \longrightarrow E'$ is proper and V is of pure dimension k, $\overline{V} \cap (E' \times U) \longrightarrow E'$ is an analytic cover, contradicting $\overline{V} \cap (\{a_{\nu_0}\} \times U) = \emptyset$. By (2.4), $V \cap (E \times U) = \emptyset$. Hence $E \subset D$, contradicting $a \in \partial D$. The claim is proved.

Define a function $\tilde{g}$ on Ω by setting $\tilde{g} = g$ on D and $\tilde{g} = 0$ on $\Omega - D$. From (2.1) we conclude that $\tilde{g}$ is a holomorphic function on Ω. Since $\tilde{g}(b) \neq 0$, $\Omega - D$ is a proper subvariety of Ω. By (2.3), $\pi^{-1}(D)$ extends to a subvariety in G. Hence $D = \Omega$, contradicting $x \in \partial D$.

Q.E.D.

(II). Theorem of Bishop

(2.7) Lemma (Federer). *Suppose X, Y are metric spaces and $f: X \longrightarrow Y$ is a map and $\lambda > 0$ such that $d(f(x), f(x')) \leq \lambda d(x, x')$ for all $x, x' \in X$ (where $d(a, a')$ is the distance between a and a'). Suppose $\delta > 0$, $\beta \geq 0$, and $c > 0$ such that, if $B \subset f(X)$ and $d(B) \leq \delta$ (where $d(B)$ is the diameter of B), then $h^\beta(B) \leq c\,d(B)^\beta$. Let $\alpha \geq 0$. If $h^{\alpha+\beta}(X) < \infty$, then*

$$\int^* h^\alpha\left(f^{-1}(y)\right) dh^\beta(y) \leq c\lambda^\beta h^{\alpha+\beta}(X),$$

where $\int^*$ denotes the upper integral.

Proof. If $A \subset X$ and $d(A) \leqq \frac{\delta}{\lambda}$, then

$$d(f(A)) \leqq \lambda d(A) \leqq \delta$$

and

$$\int^* d\left(A \cap f^{-1}(y)\right)^{\alpha} dh^{\beta}(y) \leqq d(A)^{\alpha} h^{\beta}\left(f(A)\right) \leqq c\lambda^{\beta} d(A)^{\alpha+\beta} .$$

For $0 < \epsilon \leqq \frac{\delta}{\lambda}$ and $A_\nu \subset X$ $(1 \leqq \nu < \infty)$ with $d(A_\nu) < \epsilon$ and $X \subset \bigcup_{\nu=1}^{\infty} A_\nu$, we have

$$\int^* h_\epsilon^\alpha\left(f^{-1}(y)\right) dh^\beta(y) \leqq \int^* \sum_{\nu=1}^{\infty} d\left(A_\nu \cap f^{-1}(y)\right)^{\alpha} dh^\beta(y)$$

$$\leqq \sum_{\nu=1}^{\infty} \int^* d\left(A_\nu \cap f^{-1}(y)\right)^{\alpha} dh^\beta(y)$$

$$\leqq c\lambda^\beta \sum_{\nu=1}^{\infty} d(A_\nu)^{\alpha+\beta}$$

(where, in getting the second $\leqq$, Fatou's lemma for upper integrals is used). Take inf on both sides for all admissible collection $\{A_\nu\}$. We have

$$\int^* h_\epsilon^\alpha\left(f^{-1}(y)\right) dh^\beta(y) \leqq c\lambda^\beta h_\epsilon^{\alpha+\beta}(X) .$$

Using once more Fatou's lemma for upper integrals, we obtain, as $\epsilon \longrightarrow 0$,

$$\int^* h^\alpha\left(f^{-1}(y)\right) dh^\beta(y) \leq c\lambda^\beta h^{\alpha+\beta}(X) . \qquad \text{Q.E.D.}$$

(2.8) **Lemma.** *Suppose* I *is an interval in* $\mathbb{R}$ *and* X *is a metric space. Let* $\mathbb{R} \times X$ *be given the metric such that the distance between* $(a,x) \in \mathbb{R} \times X$ *and* $(a',x') \in \mathbb{R} \times X$ *is* $\left(d(a,a')^2 + d(x,x')^2\right)^{1/2}$. *Let* $p > 0$. *Then, for any* $A \subset X$, $h^{p+1}(I \times A) \leq 2^{\frac{p+1}{2}} h^1(I)h^p(A)$.

Proof. Let $\epsilon > 0$. Choose a countable collection $\{G_\nu\}_{\nu=1}^{\infty}$ of subsets of X such that $A \subset \bigcup_{\nu=1}^{\infty} G_\nu$ and $d(G_\nu) < \epsilon$ for $1 \leq \nu < \infty$. Let E be the set of all ν such that $d(G_\nu) > 0$ and let F be the set of all ν such that $d(G_\nu) = 0$. For $\nu \in E$ cover I by a minimum number of disjoint intervals $\{I_{\nu j}\}_{j=1}^{\ell_\nu}$ of length $\leq d(G_\nu)$. We have $\ell_\nu \leq \frac{h^1(I)}{d(G_\nu)} + 1$ for $\nu \in E$. Since $p > 0$, for $\nu \in F$ we can cover I by a finite number of intervals $\{I_{\nu j}\}_{j=1}^{\ell_\nu}$ such that

$$\sum_{j=1}^{\ell_\nu} h^1(I_{\nu j})^{p+1} < \frac{\epsilon}{2^\nu}$$

and the length of each $I_{\nu j}$ is $\sqrt{2}\,\epsilon$. Clearly

$$I \times A \subset \bigcup_{\nu=1}^{\infty} \bigcup_{j=1}^{\ell_\nu} I_{\nu j} \times G_\nu .$$

We have

$$\sum_{\nu=1}^{\infty} \sum_{j=1}^{\ell_\nu} d(I_{\nu j} \times G_\nu)^{p+1} \leqq \sum_{\nu\in E} \sum_{j=1}^{\ell_\nu} 2^{\frac{p+1}{2}} d(G_\nu)^{p+1} + \sum_{\nu\in F} \sum_{j=1}^{\ell_\nu} h^1(I_{\nu j})^{p+1}$$

$$\leqq \sum_{\nu\in E} 2^{\frac{p+1}{2}} \left(h^1(I) + d(G_\nu)\right) d(G_\nu)^{\ p} + \sum_{\nu\in F} \frac{\epsilon}{2^\nu} .$$

Hence

$$h^{p+1}_{\sqrt{2}\epsilon}(I \times A) \ \leqq \ \sum_{\nu \in E} 2^{\frac{p+1}{2}} \left(h^1(I) + \epsilon\right) d(G_\nu)^p + \epsilon .$$

Taking inf on both sides for all choices of $\{G_\nu\}$, we obtain

$$h^{p+1}_{\sqrt{2}\epsilon}(I \times A) \ \leqq \ 2^{\frac{p+1}{2}} \left(h^1(I) + \epsilon\right) h^p_\epsilon(G) + \epsilon .$$

Let $\epsilon \longrightarrow 0$. It follows that

$$h^{p+1}(I \times A) \ \leqq \ 2^{\frac{p+1}{2}} \ h^1(I) h^p(A) . \qquad \text{Q.E.D.}$$

(2.9) Lemma. Suppose $\pi \colon B \longrightarrow X$ is a differentiable fiber bundle whose base and fiber are both differentiable manifolds. Let B and X be metrized by Riemannian metrics. Suppose the fiber is compact and has real dimension q . Let Q be a relatively compact open subset of X and $p > 0$. Then there exists a constant c depending only on Q and p such that $h^{p+q}(\pi^{-1}(A)) \leq c h^p(A)$ for all Borel subsets A of Q.

Proof. Cover Q by a finite number of disjoint sets W_j $(1 \leqq j \leqq k)$, each of which is the difference of two open subsets, such that $\pi^{-1}(G_j)$ is trivial for some open neighborhood G_j of W_j $(1 \leqq j \leqq k)$. Since every Borel set in X is

$S = \{\frac{1}{2} \leq |z| \leq 1\}$, $p = 2$, and k replaced by $k-1$.

Q.E.D.

(2.13) Proposition. *Suppose U is an open subset of $\mathbb{C}^n$, and P is a subvariety of U . Let A be a subvariety of $U-P$ of pure dimension k and $h^{2k}(A) < \infty$. Then $h^{2k}(\bar{A} \cap P) = 0$.*

Proof. Denote by $B(a,r)$ the open ball in $\mathbb{C}^n$ with radius r centered at a . Let K be an arbitrary compact subset of P . Let ϵ_0 be the distance of K from ∂U . Take arbitrarily $0 < \epsilon < \frac{\epsilon_0}{3n}$. Let L_ϵ be the set of all $(z_1, \ldots, z_n)$ such that $\operatorname{Re} z_j$ and $\operatorname{Im} z_j$ are integral multiples of ϵ $(1 \leq j \leq n)$. $\bigcup_{a \in L_\epsilon} B(a,n\epsilon) = \mathbb{C}^n$ and, for every $z \in \mathbb{C}^n$, the number of elements in $\{a \in L_\epsilon | z \in B(a,2n\epsilon)\}$ is at most $(4n)^{2n}$. Let I_ϵ be the set of all $a \in L_\epsilon$ such that $B(a,n\epsilon)$ intersects $\bar{A} \cap K$. Let G_ϵ be the set of points of U with distance from $K < 3n\epsilon$. Then

$$\sum_{a \in I_\epsilon} h^{2k}(A \cap B(a,2n\epsilon)) \leq (4n)^{2n} h^{2k}(A \cap G_\epsilon) .$$

Since

$$B(a,n\epsilon) \cap \bar{A} \neq \emptyset$$

for $a \in I_\epsilon$, there exists

$$b_a \in A \cap B(a,n\epsilon) .$$

By (2.12),

$$h^{2k}(A \cap B(b_a, n\epsilon)) \geq c(n\epsilon)^{2k}$$

for $a \in I_\epsilon$, where c is a positive constant depending only on n . Since $B(b_a, n\epsilon) \subset B(a, 2n\epsilon)$,

$$\sum_{a \in I_\epsilon} h^{2k}(A \cap B(a, 2n\epsilon)) \geq \sum_{a \in I_\epsilon} c(n\epsilon)^{2k} .$$

We have

$$h^{2k}_{2n\epsilon}(\bar{A} \cap K) \leq \sum_{a \in I_\epsilon} \left(\text{diameter of } B(a, n\epsilon)\right)^{2k}$$

$$= \sum_{a \in I_\epsilon} (2n\epsilon)^{2k} \leq \frac{2^{2k}}{c} (4n)^{2n} h^{2k}(A \cap G_\epsilon) .$$

Hence

$$\lim_{\epsilon \to 0} h^{2k}_{2n\epsilon}(\bar{A} \cap K) \leq \frac{2^{2k}}{c} (4n)^{2n} \lim_{\epsilon \to 0} h^{2k}(A \cap G_\epsilon) .$$

Since $h^{2k}(A) < \infty$ and for any sequence $\{\epsilon_j\}_{j=1}^{\infty}$ of positive numbers strictly decreasing to 0 the set A contains the disjoint union of $A \cap G_{\epsilon_j} - A \cap G_{\epsilon_{j+1}}$ $(1 \leq j < \infty)$, it follows that $\lim_{\epsilon \to 0} h^{2k}(A \cap G_\epsilon) = 0$. Consequently $h^{2k}(\bar{A} \cap K) = 0$ and $h^{2k}(\bar{A} \cap P) = 0$. Q.E.D.

(2.14) Theorem (Bishop). *Suppose U is an open subset of $\mathbb{C}^n$, E is a subvariety of U , and A is a subvariety of $U-E$ of pure dimension k . If $h^{2k}(A) < \infty$, then $\bar{A} \cap U$ is*

<u>a subvariety in</u> U .

<u>Proof</u>. We need only prove that $\bar{A} \cap U$ is a subvariety at an arbitrary point x of $E \cap \bar{A}$. We can assume without loss of generality that $x = 0$, U is Stein, and $B^n(1) \subset\subset U$, where $B^n(r)$ denotes the open ball in $\mathbb{C}^n$ with radius r centered at 0 . Let

$$S = \overline{B^n(1)} - B^n(\tfrac{1}{2}) .$$

By (2.10),

$$\int^*_{P \in G_{n-k}(\mathbb{C}^n)} h^1(P \cap \bar{A} \cap S) dh^{2(n-k)k}(P) \leq ch^{2k+1}(\bar{A} \cap S) ,$$

where c is a constant depending only on S . Since $h^{2k+1}(A) = 0$ and $h^{2k}(\bar{A} \cap E \cap S) = 0$ by (2.13), we have $h^{2k+1}(\bar{A} \cap S) = 0$. Hence $h^1(P \cap \bar{A} \cap S) = 0$ for almost all P in $G_{n-k}(\mathbb{C}^n)$. Fix one such P . We can assume without loss of generality that

$$P = \{z_1 = \dots = z_k = 0\} .$$

By applying (2.7) to the map from $P \cap \bar{A} \cap S$ to $\mathbb{R}$ mapping $(z_1, \dots, z_n)$ to $\left(|z_1|^2 + \dots + |z_n|^2\right)^{1/2}$, we conclude that there exists $\frac{1}{2} < r < 1$ such that

$$P \cap \bar{A} \cap S \cap \partial B^n(r) = \emptyset .$$

There exists $s > 0$ such that

$$\bar{A} \cap \left(B^k(s) \times \partial B^{n-k}(r)\right) = \emptyset$$

and

$$D := B^k(s) \times B^{n-k}(r) \subset\subset U .$$

Let $W = B^k(s)$ and let $\tilde{\pi}: \bar{A} \cap D \longrightarrow W$ be defined by

$$\tilde{\pi}(z_1, \ldots, z_n) = (z_1, \ldots, z_k) .$$

Then $\tilde{\pi}$ is proper. Let $\pi = \tilde{\pi} | A \cap D$.

Let $F = \tilde{\pi}(\bar{A} \cap E \cap D)$. F is a closed subset of measure 0 in W , because $h^{2k}(\bar{A} \cap E) = 0$ by (2.13) and π is proper. Fix $x^* \in W-F$. Since $\pi^{-1}(x^*)$ is a finite set, there exists a holomorphic function h on U which vanishes identically on E and is nowhere zero on $\pi^{-1}(x^*)$. Let G be the component of $W-F$ containing x^* . Let $g(z')$ be the holomorphic function on G which is the constant term of the polynomial $P_h(z';Z)$ in Z for the holomorphic function $h|\pi^{-1}(G)$ on the analytic cover $\pi^{-1}(G) \longrightarrow G$ as defined in the proof of (2.3). Define a function f on W by setting $f = g$ on G and $f = 0$ on $W-G$. We are going to prove that f is continuous.

Suppose, at some point $x_0 \in \partial G$, f is not continuous. Then there exist $\epsilon > 0$ and a sequence $\{x_\mu\} \subset G$ approaching x_0 such that $|f(x_\mu)| \geq \epsilon$ for all μ . Let λ be the number of sheets of the analytic cover $\pi^{-1}(G) \longrightarrow G$.

We claim that $\pi^{-1}(x_0)$ has at most λ points. Suppose the contrary. Then $\pi^{-1}(x_0)$ contains ℓ distinct points $y_1, \ldots, y_\ell$ with $\ell > \lambda$. By (2.5), $\pi^{-1}(x_0)$ is 0-dimensional. We can choose disjoint open polydisc neighborhoods $H \times V_j$ of y_j $(1 \leq j \leq \ell)$ such that $H \times V_j \subset\subset D-E$ and $(H \times \partial V_j) \cap A = \emptyset$ $(1 \leq j \leq \ell)$. Since

$(H \times V_j) \cap A \longrightarrow H$ is proper and A is of pure dimension k, $\pi((H \times V_j) \cap A) = H$ $(1 \leq j \leq \ell)$, contradicting that $\pi^{-1}(G) \longrightarrow G$ is a λ-sheeted analytic cover. The claim is proved.

Let $\pi^{-1}(x_0) = \{y_1, \ldots, y_\ell\}$, where $\ell \leq \lambda$. We claim that, for μ sufficiently large,

$$\pi^{-1}(x_\mu) \subset \bigcup_{j=1}^{\ell} H \times V_j .$$

Suppose the contrary. After we replace $\{x_\mu\}$ by a subsequence, we can assume that there exists $x_\mu' \in \pi^{-1}(x_\mu)$ such that

$$x_\mu' \notin \bigcup_{j=1}^{\ell} H \times V_j$$

and

$$x_\mu' \longrightarrow x' \in W \times \overline{B^{n-k}(r)} .$$

Since $|f(x_\mu)| \geq \epsilon$, $|h(x')| > 0$. Hence $x' \in A$ and $\pi(x') = x_0$, contradicting $x' \notin \bigcup_{j=1}^{\ell} H \times V_j$. The claim is proved.

Choose $x_\mu \in H$ such that

$$\pi^{-1}(x_\mu) \subset \bigcup_{j=1}^{\ell} H \times V_j .$$

Let

$$\varphi: \pi^{-1}(G \cap H) - \bigcup_{j=1}^{\ell} H \times \overline{V}_j \longrightarrow G \cap H$$

be induced by π . $x_\mu \notin \operatorname{Im} \varphi$. Since $(H \times \partial V_j) \cap A = \emptyset$, φ is proper. Let Q be the component of $G \cap H$ containing x_μ . Since A is of pure dimension k , it follows from $x_\mu \notin \operatorname{Im} \varphi$ that $Q \cap \operatorname{Im} \varphi = \emptyset$. The set $\pi^{-1}(H) - \bigcup_{j=1}^{\ell} H \times \overline{V}_j$ is empty, otherwise it is a subvariety of pure dimension k in

$$H \times \left(B^{n-k}(r) - \bigcup_{j=1}^{\ell} \overline{V}_j\right) - E$$

and is disjoint from both

$$H \times \left(\partial B^{n-k}(r) \cup \left(\bigcup_{j=1}^{\ell} V_j\right)\right)$$

and $Q \times B^{n-k}(r)$, contradicting (2.4). The continuity of f follows, because the emptiness of

$$\pi^{-1}(H) - \bigcup_{j=1}^{\ell} H \times \overline{V}_j$$

contradicts $x_0 \in F$.

From (2.1) we conclude that f is holomorphic on W . Since $f(x^*) \neq 0$, $W-G$ is contained in a proper subvariety of W . By (2.3), $\pi^{-1}(G)$ can be extended to a subvariety in D . $\overline{A} \cap U$ is a subvariety at 0 . Q.E.D.

(2.15) Theorem (Shiffman). *Suppose* U *is an open subset in* $\mathbb{C}^n$, E *is a closed subset of* U *with* $h^{2k-1}(E) = 0$, *and* A *is a subvariety of pure dimension* k *in* $U-E$. *Then* $\overline{A} \cap U$ *is a subvariety in* U .

Proof. We need only prove that $\bar{A} \cap U$ is a subvariety at an arbitrary point x of $E \cap \bar{A}$. We can assume without loss of generality that $x = 0$. Since $h^{2k-1}(E) = 0$, $h^{2k+1}(\bar{A}) = 0$. As in the first paragraph of the proof of (2.14), after a linear coordinates transformation we can assume that there exist $s > 0$ and $r > 0$ such that

$$\bar{A} \cap \left(B^k(s) \times \partial B^{n-k}(r)\right) = \emptyset$$

and

$$D := B^k(s) \times B^{n-k}(r) \subset\subset U .$$

Let $W = B^k(s)$, let $\pi: \bar{A} \cap D \longrightarrow W$ be defined by

$$\pi(z_1, \ldots, z_n) = (z_1, \ldots, z_k) ,$$

and $F = \pi(\bar{A} \cap E \cap D)$. $h^{2k-1}(F) = 0$. By the lemma below, $W-F$ is connected. Since $\pi^{-1}(W-F)$ is an analytic cover over $W-F$, the theorem follows from (2.3) and the lemma below.

Q.E.D.

(2.16) Lemma. Suppose U is an open subset of $\mathbb{C}^n$ and E is a closed subset of U with $h^{2n-1}(E) = 0$. Then any holomorphic function f on $U-E$ which is locally bounded on U can be extended to a holomorphic function on U.

Proof. We need only show that f can be extended to a holomorphic function in an open neighborhood of an arbitrary point x of E. We can assume without loss of generality that $x = 0$. Since $h^{2n-1}(E) = 0$, as in the first paragraph of the proof of (2.14), after a linear coordinates transformation we conclude that there exist $s > 0$ and $r > 0$ such that

$$E \cap (B^{n-1}(s) \times \partial B^1(r)) = \emptyset$$

and

$$B^{n-1}(s) \times B^1(r) \subset\subset U .$$

First we consider the case $n = 1$. Let α be the distance between $E \cap B^1(r)$ and $\partial B^1(r)$. Fix arbitrarily $\eta > 0$. Since $h^1(E \cap B^1(r)) = 0$, there exist a finite number of open discs G_i $(1 \leq i \leq k)$ such that $E \cap B^1(r) \cap G_i \neq \emptyset$, $E \cap B^1(r) \subset \bigcup_{i=1}^{k} G_i$, diameter of $G_i < \frac{\alpha}{2}$, and

$\sum_{i=1}^{k}$ diameter of $G_i < \eta$. Let $\gamma_\eta = \partial\left(\bigcup_{i=1}^{k} G_i\right)$. The length of

γ_η is $< \pi\eta$. For $z \in B^1(r) - \bigcup_{i=1}^{k} \overline{G}_i$,

$$f(z) = \frac{1}{2\pi i} \int_{\partial B^1(r)} \frac{f(\zeta)d\zeta}{\zeta - z} - \frac{1}{2\pi i} \int_{\gamma_\eta} \frac{f(\zeta)d\zeta}{\zeta - z} .$$

Since f is locally bounded in U, by letting $\eta \longrightarrow 0$, we conclude that

$$f(z) = \frac{1}{2\pi i} \int_{\partial B^1(r)} \frac{f(\zeta)d\zeta}{\zeta - z} .$$

Hence f can be extended to a holomorphic function on $B^1(r)$.

For the general case, by applying (2.7), with

$$X = E \cap (B^{n-1}(s) \times B^1(r)) ,$$

$Y = B^{n-1}(s)$, and $X \to Y$ induced by the natural projection $B^{n-1}(s) \times B^1(r) \longrightarrow B^{n-1}(s)$, we conclude from $h^{2n-1}(E) = 0$ that there exists a set A of measure 0 in $B^{n-1}(s)$ such that

$$h^1\left(E \cap \left(\{x\} \times B^1(r)\right)\right) = 0$$

for $x \in B^{n-1}(s) - A$. By the special case of $n = 1$, we know that, for $x \in B^{n-1}(s) - A$, $f|\left(\{x\} \times B^1(r)\right) - E$ can be extended to a holomorphic function on $\{x\} \times B^1(r)$. By (1.2) f can be extended to a holomorphic function on $B^{n-1}(s) \times B^1(r)$. Q.E.D.

III. Theorem of Rothstein

First we prove a lemma needed for the uniqueness part of Rothstein's theorem.

(2.17) Lemma. Suppose G is an open subset of $\mathbb{C}^N$, K is a compact subset of G , D is an open subset of $\mathbb{C}^n$, and E is a closed subset of $D \times G$. Suppose one of the following two conditions (i), (ii) is satisfied:

(i) There exist holomorphic functions f_i on $D \times G$ ($i \in I$) such that

$$E = \{x \in D \times G \mid |f_i(x)| \leq 1 \text{ for all } i \in I\} .$$

(ii) There exist holomorphic functions g_j on $D \times G$ ($j \in J$) such that

$$E = \{x \in D \times G \mid \mathrm{Re}\, g_j(x) = 0 \text{ for all } j \in J\} .$$

Then the following conclusions hold:

(a) If A is a subvariety in $(D \times G)-E$ such that $A \subset D \times K$, then $\dim A \leq n$.

(b) If A_i is a subvariety in $(D \times G)-E$ whose every branch has dimension $\geq n+1$ $(i = 1,2)$ such that

$$A_1 \cap (D \times (G-K)) = A_2 \cap (D \times (G-K)) ,$$

then $A_1 = A_2$.

Proof. Condition (i) follows from condition (ii) by setting

$$\{f_i\}_{i \in I} = \{e^{g_j}, e^{-g_j}\}_{j \in J} .$$

So we can assume that we have condition (i).

(a) By considering the subvariety $A \cap (\{t\} \times G)$ of $(\{t\} \times G)-E$ for every $t \in D$, we can reduce the general case to the special case $n = 0$.

Suppose $\dim A > n = 0$. We are going to derive a contradiction. We can assume that A is irreducible. Take $x \in A$. Then $|f_i(x)| > 1$ for some $i \in I$. The sup of $|f_i|$ on A is assumed at some point of A, because the sup of $|f_i|$ on A equals the sup of $|f_i|$ on the compact set

$$A \cap \{|f_i| \geq |f_i(x)|\} .$$

By the maximum modulus principle, $f_i \equiv f_i(x)$ on A. It follows that A is compact, contradicting $\dim A > 0$.

(b) Let A_1' be a branch of A. By (a),

$$A_1' \cap \left(D \times (G-K)\right) \neq \emptyset .$$

Hence a nonempty open subset of A_1' is contained in A_2. It follows that $A_1' \subset A_2$. Likewise every branch A_2' of A_2 is contained in A_1. Consequently $A_1 = A_2$. Q.E.D.

We introduce the following notations for the rest of these lecture notes: For $a \in \mathbb{R}^N$, we denote by $a_1, \ldots, a_N$ the coordinates of a. For $a, b \in \mathbb{R}^N$ we say that $a < b$ (respectively $a \leqq b$) if $a_i < b_i$ for $1 \leqq i \leqq N$ (respectively $a_i \leqq b_i$ for $1 \leqq i \leqq N$). For $0 < b$ in $\mathbb{R}^N$ denote by $\Delta^N(b)$ the polydisc

$$\{(z_1, \ldots, z_N) \in \mathbb{C}^N \,\big|\, |z_i| < b_i \text{ for } 1 \leqq i \leqq N\} .$$

For $0 \leqq a < b$ denote by $G^N(a,b)$ the set

$$\{(z_1, \ldots, z_N) \in \Delta^N(b) \,\big|\, |z_i| > a_i \text{ for some } 1 \leqq i \leqq N\} .$$

When $a_1 = \ldots = a_N = r$ and $b_1 = \ldots = b_N = s$, we write $\Delta^N(s)$, $G^N(r,s)$ instead of $\Delta^N(a)$, $G^N(a,b)$. When $N = 1$, $\Delta^N(s)$ is simply denoted by $\Delta(s)$; and $\Delta^N(1)$ is simply denoted by Δ. These agree with our earlier notations.

(2.18) Theorem. *Suppose D is a connected open subset of $\mathbb{C}^n$. Suppose $0 \leqq a < b$ in $\mathbb{R}^N$ and V is a subvariety of $D \times G^N(a,b)$ whose every branch has dimension $\geq n+1$. Suppose A is a thick set in D such that, for every $t \in A$,*

$V \cap (\{t\} \times G^N(a,b))$ <u>can be extended to a subvariety in</u> $\{t\} \times \Delta^N(b)$. <u>Then</u> V <u>can be extended uniquely to a subvariety</u> $\tilde{V}$ <u>in</u> $D \times \Delta^N(b)$ <u>such that every branch of</u> $\tilde{V}$ <u>has dimension</u> $\geq n+1$.

<u>Proof</u>. The uniqueness of $\tilde{V}$ follows from (2.17)(b).

To prove the existence of $\tilde{V}$, we introduce the following notations: Suppose E is a subset of $\mathbb{C}^n \times \mathbb{C}^N$, H is a subset of $\mathbb{C}^n$, and $0 \leq c < d$ in $\mathbb{R}^N$. $E(H)$ denotes $E \cap (H \times \mathbb{C}^N)$. When $H = \{t\}$, we write $E(t)$ instead of $E(\{t\})$. $E_{(c,d)}$, $E_{[c,d]}$, and $E_{\overline{[c,d)}}$ denote respectively $E \cap (\mathbb{C}^n \times G^N(c,d))$, $E \cap (\mathbb{C}^n \times \overline{G^N(c,d)})$, and $E \cap (\mathbb{C}^n \times (\overline{G^N(c,d)} \cap \Delta^N(d)))$.

Let A' be the set of all $t \in A$ such that, for every open neighborhood U of t in D , $U \cap A$ is thick. It is clear that A' is thick.

(a) We make the following additional assumptions:

$$\begin{cases} V \text{ is of pure dimension } d . \\ \dim V(t) \leq d-n \text{ for } t \in A' . \end{cases}$$

We are going to prove that for every $t \in A'$ there exists an open neighborhood W of t in D such that $V(W)$ can be extended to a subvariety in $W \times \Delta^N(b)$. For $t \in A'$ let $V(t)^\sim$ be the pure-dimensional subvariety in $\{t\} \times \Delta^N(b)$ which extends $V(t)$. Fix $t^0 \in A'$. Take $a < a' < b' < b$ in $\mathbb{R}^N$. If $V(t_0) = \emptyset$, then $W \cap V_{[a',b']} = \emptyset$ for some open neighborhood W of t in D , and, by (2.17)(a),

$W \cap V_{(a,b']} = \emptyset$, which implies that $V(W)$ is a subvariety of $W \times \Delta^N(b)$. Hence we can assume that $V(t_0)$ has pure dimension $d-n$. By the theorem on the existence of special analytic polyhedra (see the Appendix of Chapter 2) there exist holomorphic functions $f_1, \ldots, f_k$ on $\mathbb{C}^n \times \mathbb{C}^N$ (where $k = d-n$) and an open neighborhood U of $\{t^0\} \times \overline{\Delta^N(a')}$ in $D \times \Delta^N(b')$ such that

$$V(t^0)^{\sim}_{[0,a']} \subset U \cap V(t^0)^{\sim} \cap F^{-1}(\Delta^k) \subset\subset U \cap V(t^0)^{\sim} ,$$

where $F: \mathbb{C}^n \times \mathbb{C}^N \longrightarrow \mathbb{C}^k$ is defined by $f_1, \ldots, f_k$. There exists a relatively compact open neighborhood B of $\{t^0\} \times \overline{\Delta^N(a')}$ in U with

$$B \cap V(t^0)^{\sim} = U \cap V(t^0)^{\sim} \cap F^{-1}(\Delta^k).$$

Choose $0 < \alpha < 1$ such that

$$V(t^0)^{\sim}_{[0,a']} \subset U \cap V(t^0)^{\sim} \cap F^{-1}(\Delta^k(\alpha)) .$$

Take $\alpha < \beta < 1$. There exists an open neighborhood W' of t^0 in D such that

(i) $W' \times \overline{\Delta^N(a')} \subset B$,

(ii) $V \cap (W' \times \partial(\Delta^N(a'))) \subset F^{-1}(\Delta^k(\alpha))$,

(iii) $U \cap V_{[a',b)} \cap F^{-1}(\overline{\Delta^k(\beta)})$ is disjoint from $(\partial B)(W')$.

The map φ from

$$X: = V \cap B(W')_{(a',b')} \cap F^{-1}(G^k(\alpha,\beta))$$

to $W' \times G^k(\alpha,\beta)$ defined by $(t_1, \ldots, t_n, f_1, \ldots, f_k)$

(where $t_1, \ldots, t_n$ are the coordinates of $\mathbb{C}^n$) is proper, because, if K is a compact subset of W' and $\alpha < \alpha' < \beta' < \beta$, then the inverse image of $K \times \overline{G^k(\alpha',\beta')}$ under φ is

$$\overline{B} \cap V_{[a',b']}(K) \cap F^{-1}\left(\overline{G^k(\alpha,\beta)}\right)$$

which is compact. For $t \in W' \cap A'$ the map ψ_t from

$$V(t)^*: \; = \; B \cap V(t)^{\sim} \cap F^{-1}(\Delta^k(\beta))$$

to $\{t\} \times \Delta^k(\beta)$ is proper, because, if $0 < \beta' < \beta$, then the inverse image of $\{t\} \times \overline{\Delta^k(\beta')}$ under ψ_t is

$$\overline{B} \cap V(t)^{\sim} \cap F^{-1}(\overline{\Delta^k(\beta')})$$

which is compact. By (2.3)(B)(b)(c) applied to the analytic cover

$$\varphi: \; X \longrightarrow W' \times G^k(\alpha,\beta) ,$$

X can be extended to a subvariety $\tilde{X}$ in

$$\left(W' \times \Delta^N(b')\right) \cap F^{-1}\left(\Delta^k(\beta)\right)$$

such that the map

$$\tilde{X} \longrightarrow W' \times \Delta^k(\beta)$$

defined by $(t_1, \ldots, t_n, f_1, \ldots, f_k)$ makes $\tilde{X}$ an analytic cover over $W' \times \Delta^k(\beta)$. By (2.3)(A), $\tilde{X}(t^0) = V(t^0)^{\sim}$. Hence

$$\tilde{X}(t^0)_{[0,a']} \subset F^{-1}(\Delta^k(\alpha)) .$$

Take $\alpha < \beta^* < \beta$. There exists an open neighborhood W'' of t^0 in W such that

$$(\dagger) \qquad \tilde{X}(W'')_{[0,a']} \cap F^{-1}(\overline{\Delta^k(\beta^*)}) \subset F^{-1}(\Delta^k(\alpha)) .$$

Since $\tilde{X} \cap F^{-1}(\overline{\Delta^k(\beta^*)})$ is disjoint from $(\partial B)(t^0)$, there exists an open neighborhood W of t^0 in W'' such that

$$(\dagger\dagger) \qquad \tilde{X} \cap F^{-1}(\overline{\Delta^k(\beta^*)}) \text{ is disjoint from } (\partial B)(W) .$$

Let V' be the union of

$$\tilde{X} \cap B(W) \cap F^{-1}(\Delta^k(\beta^*))$$

and

$$V \cap B(W)_{(a',b)} \cap F^{-1}(\mathbb{C}^k - \overline{\Delta^k(\alpha)}) .$$

Being the union of two locally closed subvarieties in $B(W)$, V' is a locally closed subvariety in $B(W)$. Take $\alpha < \alpha' < \beta' < \beta^*$. Since X is the intersection of $\tilde{X}$ and

$$B(W')_{(a',b')} \cap F^{-1}(G^k(\alpha,\beta)) ,$$

it follows from (ii) and ($\dagger$) that V' is the union of

$$\tilde{X} \cap B(W) \cap F^{-1}(\overline{\Delta^k(\beta')})$$

and

$$V \cap B(W)_{[a',b)} \cap F^{-1}(\mathbb{C}^k - \Delta^k(\alpha'))$$

which are both closed subsets of $B(W)$. Hence V' is a subvariety of $B(W)$.

Let

$$V^* = V' \cup \big((V-B)(W)\big) .$$

We claim that V^* is a subvariety of $W \times \Delta^N(b)$. Take arbitrarily $x \in (\partial B)(W)$. Because of (i), (iii), and (††), we can choose an open neighborhood Q of x in U which is disjoint from $W \times \overline{\Delta^N(a')}$,

$$U \cap V_{[a',b)} \cap F^{-1}\big(\overline{\Delta^k(\beta)}\big) ,$$

and $\tilde{X} \cap F^{-1}\big(\overline{\Delta^k(\beta^*)}\big)$. Then $Q \cap V^* = Q \cap V$. It follows that V^* is a subvariety of $W \times \Delta^N(b)$. The claim is proved. Since $V^*_{(b',b)} = V(W)_{(b',b)}$, by (2.17)(b) $V^*_{(a,b)} = V(W)$. Hence we have proved that for every $t \in A'$ there exists an open neighborhood W of t in D such that $V(W)$ can be extended to a subvariety in $W \times \Delta^N(b)$.

(b) Let D' be the largest open subset of D such that $V(D')$ can be extended to a subvariety in $D' \times \Delta^N(b)$. It follows from (a) that, under the following additional assumptions:

$$\begin{cases} V \text{ is of pure dimension } d . \\ \dim V(t) \leq d-n \text{ for } t \in D , \end{cases}$$

D' is a nonempty closed subset of D . Hence $D' = D$.

(c) Let $\pi: V \longrightarrow D$ be induced by the natural projection $D \times G^N(a,b) \longrightarrow D$. Let S be the closure of the set of points of V where the rank of π is $< n$. Take $a < a' < b' < b$. Then $\pi\big(S_{[a',b']}\big)$ is a closed thin set in

D . Let $D' = D - \pi(S_{[a',b']})$.

Let $V = \bigcup_{i \in I} V^{(i)}$ be the decomposition of V into pure-dimensional components. Let C be the set of all $t \in D$ such that, for some $i \neq j$, some nonempty open subset of $V^{(i)}(t)$ is contained in $V^{(j)}$. Then C is thin in D (see the Appendix of Chapter 2 for theorems on thick sets). For $t \in A-C$, $V^{(i)}(t)$ can be extended to a subvariety of $\{t\} \times \Delta^N(b)$ for $i \in I$. By applying (b) to the subvariety $V^{(i)}(D')_{(a',b')}$ of $D' \times G^N(a',b')$ and to the thick set $A-\pi(S_{[a',b']})-C$ in D' , we conclude that $V^{(i)}(D')_{(a',b')}$ can be extended to a subvariety in $D' \times \Delta^N(b')$ for $i \in I$. By (2.17)(b), $V^{(i)}(D')$ can be extended to a subvariety in $D' \times \Delta^N(b)$ for $i \in I$.

Let L be an arbitrary relatively compact open subset of D . By (2.2) there exists an N-dimensional plane T in $\mathbb{C}^n \times \mathbb{C}^N$ such that for some nonempty open subset Q of D' and for some open neighborhood R of L in D we have

(i) $(Q+T) \cap (\mathbb{C}^n \times \mathbb{C}^N(b)) \subset D' \times \Delta^N(b)$

(ii) $L \times \Delta^N(b) \subset (R+T) \cap (\mathbb{C}^n \times \Delta^N(b)) \subset D \times \Delta^N(b)$

(iii) $\dim(x+T) \cap V^{(i)} \leq \dim V^{(i)} - n$
for $x \in V^{(i)}$ and $i \in I$,

where

$$Q + T = \{x+y \mid x \in Q,\ y \in T\}$$

and

$$x + T = \{x+y \mid y \in T\} .$$

By (b) there exists a subvariety $\tilde{V}^{(i)}$ in

$$(R+T) \cap (\mathbb{C}^n \times \Delta^N(b))$$

such that

$$\tilde{V}^{(i)} \cap (R+T) \cap (\mathbb{C}^n \times G^N(a,b)) = V^{(i)} \cap (R+T) \cap (\mathbb{C}^n \times G^N(a,b))$$

$(i \in I)$. $\tilde{V}^{(i)} \cap (L \times \Delta^N(b))$ is a subvariety in $L \times \Delta^N(b)$ extending $V^{(i)}(L)$ for $i \in I$. Since L is an arbitrary relatively compact open subset of D , V can be extended to a subvariety in $D \times \Delta^N(b)$. Q.E.D.

(2.19) Corollary. (Rothstein). *Suppose* $0 \leq a < b$ *in* $\mathbb{R}^N$, D *is a connected open subset of* $\mathbb{C}^n$, D' *is a nonempty open subset of* D , *and* V *is a subvariety in* $(D \times G^N(a,b)) \cup (D' \times \Delta^N(b))$ *whose every branch has dimension* $\geq n+1$. *Then* V *can be extended uniquely to a subvariety in* $D \times \Delta^N(b)$ *whose every branch has dimension* $\geq n+1$.

IV. Extension across $\mathbb{R}^n$

We will distinguish between the case of a subvariety whose every branch has dimension at least two and the case of a subvariety of pure dimension one. We will prove the first case from Rothstein's theorem and the second case by projections defined by quadratic polynomials. However, the first case can also be derived in the same way as the second case.

(2.20) Lemma. If U is an open neighborhood of 0 in $\mathbb{C}^n$ $(n \geqq 2)$, then there exists a biholomorphic map σ from Δ^n onto an open neighborhood of 0 in U such that the image of $(\Delta(\frac{1}{2}) \times \Delta^{n-1}) \cup (\Delta \times G^{n-1}(\frac{1}{2},1))$ under σ is disjoint from $\mathbb{R}^n$.

Proof. Let $\alpha: \mathbb{C}^n \longrightarrow \mathbb{C}^n$ be defined by

$$(z_1, \ldots, z_n) \longrightarrow (e^{iz_1}, \ldots, e^{iz_n}) .$$

The image of $\mathbb{R}^n$ under α is $(\partial\Delta)^n$ and α maps an open neighborhood D of 0 in U biholomorphically onto an open neighborhood W of $x_0 := (1, \ldots, 1)$ in $\mathbb{C}^n$.

Let L be the hyperplane in $\mathbb{C}^n$ defined by

$$z_1 + \ldots + z_n = n .$$

Since

$$\mathrm{Re}\, z_1 + \ldots + \mathrm{Re}\, z_n - n = 0$$

on L, we have

$$L \cap (\partial\Delta)^n = \{x_0\} .$$

Let w_j be a linear function of $z_1, \ldots, z_n$ $(1 \leqq j \leqq n)$ such that

(i) $\left(\dfrac{\partial w_j}{\partial z_k}\right)_{1 \leqq j,k \leqq n}$ is nonsingular,

(ii) $w_1 = c\,(z_1 + \ldots + z_n - n)$, where $c > 0$,

(iii) $w_j(x_0) = 0$ $(1 \leqq j \leqq k)$,

(iv) $\{x \in \mathbb{C}^n \mid |w_j(x)| \leq 1,\ 1 \leq j \leq k\} \subset\subset W$.

Since $L \cap (\partial\Delta)^n = \{x_0\}$, there exists $0 < \eta < 1$ such that

$$\{x \in \mathbb{C}^n \mid |w_j(x)| \leq \eta,\ \tfrac{1}{2} \leq \max_{2 \leq j \leq n} |w_j(x)| \leq 1\}$$

is disjoint from $(\partial\Delta)^n$.

Let $\beta: \mathbb{C}^n \longrightarrow \mathbb{C}^n$ be defined by

$$\beta(x) = \left(\frac{2w_1(x)}{\eta} - \frac{1}{2},\ w_2(x),\ \ldots,\ w_n(x)\right).$$

Since

$$\{x \in \mathbb{C}^n \mid \operatorname{Re} w_1(w) > 0\}$$

is disjoint from $(\partial\Delta)^n$, the map $\sigma = ((\beta \circ \alpha)|D)^{-1}|\Delta^n$ satisfies the requirement. Q.E.D.

(2.21) <u>Theorem</u>. <u>Suppose</u> G <u>is an open subset of</u> $\mathbb{C}^n$ <u>and</u> X <u>is a subvariety of</u> $G - \mathbb{R}^n$ <u>whose every branch has dimension</u> ≥ 2 . <u>Then</u> $G \cap \overline{X}$ <u>is a subvariety of</u> G .

<u>Proof</u>. By (2.20), (2.19), and (2.17)(b) for every point $x \in \mathbb{R}^n$ there exists an open neighborhood U of x in G such that $X \cap U$ can be extended to a subvariety $\tilde{X}$ of U . We can assume without loss of generality that no branch of $\tilde{X}$ is contained in $\mathbb{R}^n$. Since $\operatorname{Re} z_j$ is a real-analytic function $(1 \leq j \leq n)$, $\tilde{X} \cap \mathbb{R}^n$ is nowhere dense in X . It follows that $G \cap \overline{X}$ is a subvariety of G . Q.E.D.

A subvariety X of an open subset of $\mathbb{C}^n$ is said to be self-conjugate if $(\bar{z}_1, \ldots, \bar{z}_n) \in X$ whenever $(z_1, \ldots, z_n) \in X$.

(2.22) Theorem (Alexander). *Suppose G is an open subset of $\mathbb{C}^n$ and X is a self-conjugate subvariety of $G-\mathbb{R}^n$ of pure dimension 1. Then $G \cap \overline{X}$ is a subvariety in G.*

Proof. We say that a self-conjugate subvariety of an open subset of $\mathbb{C}^n$ is an irreducible self-conjugate subvariety if it is not the union of two of its proper self-conjugate subvarieties. First we show that,

(*) $\left\{\begin{array}{l}\text{if } V \text{ is an irreducible self-conjugate subvariety} \\ \text{of pure dimension } 1 \text{, then } V-\mathbb{R}^n \text{ is an irredu-} \\ \text{cible self-conjugate subvariety.}\end{array}\right.$

Suppose the contrary. Then $V-\mathbb{R}^n = V_1 \cup V_2$, where V_i is a proper self-conjugate subvariety of $V-\mathbb{R}^n$ $(i = 1,2)$ and $\dim V_1 \cap V_2 \leqq 0$. Let S = singular set of V. Since $\mathbb{R}^n \cap V$ is nowhere dense in V (otherwise $\mathbb{R}^n$ contains the tangent space of a regular point of V which is a complex line) and since $V-S$ is connected, some point x of $\mathbb{R}^n \cap (V-S)$ belongs to $\overline{V}_1 \cap \overline{V}_2$. The intersection of $\mathbb{R}^n$ with the tangent space T of V at x has real dimension 1, because

$$T = (\mathbb{R}^n \cap T) + i(\mathbb{R}^n \cap T) .$$

Hence after a linear transformation of coordinates in $\mathbb{C}^n$ with real coefficients, we can assume without loss of generality that $x = 0$ and $\pi(\mathbb{R}^n \cap T)$ has real dimension 1, where $\pi: \mathbb{C}^n \longrightarrow \mathbb{C}$ is defined by $\pi(z_1, \ldots, z_n) = z_1$. Since T is a complex line, $\pi(T) = \mathbb{C}$. It follows that π maps an open neighborhood H of 0 in $V-S$ biholomorphically onto $\Delta(\epsilon)$ for some $\epsilon > 0$. Being biholomorphic to $\Delta(\epsilon)-\mathbb{R}$, $H-\pi^{-1}(\mathbb{R})$ has only two components and these two components are conjugate to each other. Since $H-\pi^{-1}(\mathbb{R})$ is the union of $H \cap V_1 - \pi^{-1}(\mathbb{R})$ and $H \cap V_2 - \pi^{-1}(\mathbb{R})$, the two components of $H-\pi^{-1}(\mathbb{R})$ are precisely $H \cap V_1 - \pi^{-1}(\mathbb{R})$ and $H \cap V_2 - \pi^{-1}(\mathbb{R})$, contradicting that V_1 and V_2 are both self-conjugate and are disjoint from each other. (*) is proved.

In view of (*), to prove the theorem, we need only show that there exists an open neighborhood U of 0 in G such that $X \cap U$ is contained in a subvariety of U of dimension 1.

Let $\sigma: \mathbb{C}^n \longrightarrow \mathbb{C}$ be defined by

$$\sigma(z_1, \ldots, z_n) = (z_1^2, \ldots, z_n^2).$$

By replacing G by an open polydisc neighborhood G' of 0 in G and replacing X by $\sigma(X \cap G')$, we can assume without loss of generality that $\overline{X} \cap (G-\mathbb{R}_+^n)$ is a subvariety of $G-\mathbb{R}_+^n$, where $\mathbb{R}_+$ is the set of all nonnegative real numbers. Choose $\lambda_1, \ldots, \lambda_n > 0$ such that $\lambda_1 z_1 + \ldots + \lambda_n z_n$ does not vanish identically on any branch of $\overline{X} \cap (G-\mathbb{R}_+^n)$. Let $\pi: \mathbb{C}^n \longrightarrow \mathbb{C}$ be defined by

$$\pi(z_1, \ldots, z_n) = \lambda_1 z_1 + \ldots + \lambda_n z_n.$$

Since

$$\pi^{-1}(0) \cap \mathbb{R}_+^n = \{0\} ,$$

$\pi^{-1}(0) \cap \overline{X}$ is countable. After a homogeneous linear coordinates transformation of $\mathbb{C}^n$ with coefficients in $\mathbb{R}_+$, we can assume without loss of generality that $\pi(z_1, \ldots, z_n) = z_1$. There exists $0 < r < 1$ such that

$$\overline{X} \cap \left(\overline{\Delta(r)} \times \partial(\Delta^{n-1}(r))\right) = \emptyset$$

and $\overline{\Delta^n(r)} \subset G$.

$$X' := X \cap \left((\Delta(r) - \mathbb{R}) \times \Delta^{n-1}(r)\right)$$

is an analytic cover over $\Delta(r) - \mathbb{R}$. Let its critical set be A and its number of sheets by λ.

Fix $2 \leqq j \leqq n$. Let

$$P_j(z_1;Z) = Z^\lambda + \sum_{i=0}^{\lambda-1} a_i(z_1)Z^i$$

be the polynomial associated to the holomorphic function $z_j|X'$ on the analytic cover $X' \longrightarrow \Delta(r) - \mathbb{R}$. We are going to show that $\operatorname{Im} a_i(z_1) \longrightarrow 0$ as $z_1 \longrightarrow$ some point of $\mathbb{R} \cap \Delta(r)$. Fix $1 \leqq i \leqq \lambda-1$. Suppose $z_1^{(\nu)} \in \Delta(r) - \mathbb{R}$ $(\nu \geqq 1)$ such that $z_1^{(\nu)} \longrightarrow z_1^* \in \mathbb{R} \cap \Delta(r)$ as $\nu \longrightarrow \infty$. Take

$$\zeta^{(\nu)} \in \Delta(r) - \mathbb{R} - A$$

such that

$$\left|a_i(z_1^{(\nu)}) - a_i(\zeta^{(\nu)})\right| < \frac{1}{\nu}$$

and $\left|z_1^{(\nu)} - \zeta^{(\nu)}\right| < \frac{1}{\nu}$. It suffices to show that Im a_i approaches 0 on some subsequence of $\{\zeta^{(\nu)}\}$.

Let

$$X' \cap \pi^{-1}(\zeta^{(\nu)}) = \{\zeta_1^{(\nu)}, \ldots, \zeta_\lambda^{(\nu)}\} .$$

By replacing $\{\zeta^{(\nu)}\}$ by a subsequence, we can assume that $\zeta_k^{(\nu)} \longrightarrow \zeta_k^*$ as $\nu \longrightarrow \infty$ $(1 \leq k \leq \lambda)$. Clearly $\pi(\zeta_k^*) = z_1^*$ $(1 \leq k \leq \lambda)$. For $\epsilon > 0$ let T_ϵ be the set of all $(z_1, \ldots, z_n) \in \mathbb{C}^n$ such that $|\text{Im } z_\ell| < \epsilon$ $(1 \leq \ell \leq n)$. Fix ϵ . After renumbering $\zeta_1^*, \ldots, \zeta_\lambda^*$, we can assume that $\zeta_1^*, \ldots, \zeta_m^*$ are not in T_ϵ and $\zeta_{m+1}^*, \ldots, \zeta_\lambda^*$ are in T_ϵ .

We claim that

$$X' \cap \pi^{-1}(z_1^*) - T_\epsilon = \{\zeta_1^*, \ldots, \zeta_m^*\} .$$

Suppose the contrary. There exists

$$\tilde{\zeta} \in X' \cap \pi^{-1}(z_1^*) - T_\epsilon - \{\zeta_1^*, \ldots, \zeta_m^*\} .$$

By (2.17) $X' \cap \pi^{-1}(z_1^*)$ is at most 0-dimensional. Hence there exists a relatively compact open neighborhood $D_1 \times D_2$ of $\tilde{\zeta}$ in

$$\Delta^n(r) - T_\epsilon - \{\zeta_1^*, \ldots, \zeta_m^*\}$$

such that D_1 is an open subset of $\mathbb{C}$, D_2 is an open subset of $\mathbb{C}^{n-1}$, and $D_1 \times \partial D_2$ is disjoint from X . Since $X \cap (D_1 \times D_2) \longrightarrow D_1$ induced by π is proper and X has pure dimension 1 ,

$$\pi(X \cap (D_1 \times D_2)) = D_1 .$$

For ν sufficiently large, $\zeta^{(\nu)} \in D_1$ and $\zeta_k^{(\nu)} \notin D_1 \times D_2$ for $1 \leq k \leq \lambda$, contradicting

$$X' \cap \pi^{-1}(\zeta^{(\nu)}) = \{\zeta_1^{(\nu)}, \ldots, \zeta_\lambda^{(\nu)}\} .$$

The claim is proved.

Since $z_1^* \in \mathbb{R}$ and $X' - T_\epsilon$ is self-conjugate, the conjugate of $\{\zeta_1^*, \ldots, \zeta_m^*\}$ is itself. The coefficients of

$$R(Z): = \prod_{k=1}^{m} \left(Z - z_j(\zeta_k^*)\right)$$

are all real and have absolute values $\leq 2^\lambda$. Since $\zeta_k^* \in T_\epsilon$ for $m < k \leq \lambda$, the imaginary parts of the coefficients of

$$S(Z): = \prod_{k=m+1}^{\lambda} \left(Z - z_j(\zeta_k^*)\right)$$

have absolute values $\leq 4^\lambda \epsilon$. Let

$$R(Z)S(Z) = Z^\lambda + \sum_{i=0}^{\lambda-1} a_i^* Z^i .$$

Then $|\operatorname{Im} a_i^*| < \lambda 8^\lambda \epsilon$ $(0 \leq i < \lambda)$. Since $a_i(\zeta^{(\nu)}) \longrightarrow a_i^*$ $(0 \leq i < \lambda)$, $|\operatorname{Im} a_i(\zeta^{(\nu)})| < 2\lambda 8^\lambda \epsilon$ for ν sufficiently large $(0 \leq i < \lambda)$. It follows that $\operatorname{Im} a_i(\zeta^{(\nu)}) \longrightarrow 0$ as $\nu \longrightarrow \infty$ for $0 \leq i < \lambda$. So $\operatorname{Im} a_i(z_1) \longrightarrow 0$ as $z_1 \longrightarrow$ some point of $\mathbb{R} \cap \Delta(r)$ $(0 \leq i < \lambda)$.

Since X' is self-conjugate, $a_i(\bar{z}_1) = \overline{a_i(z_1)}$

$(0 \leq i < \lambda)$. It follows that $a_i(z_1) - a_i(\bar{z}_1) \longrightarrow 0$ as $z_1 \longrightarrow z_1^* \in \mathbb{R} \cap \Delta(r)$ and the convergence is uniform for z_1^* lying in a fixed compact subset of $\mathbb{R} \cap \Delta(r)$. Since $a_i(z_1)$ is uniformly bounded on $\Delta(r) - \mathbb{R}$, for $0 < s < r$ and $z_1 \in \Delta(s) - \mathbb{R}$ we have

$$\frac{1}{2\pi i} \int_{\zeta \in (\partial\Delta(s)) - \mathbb{R}} \frac{a_i(\zeta)d\zeta}{\zeta - z_1} = \lim_{\epsilon \to 0^+} \frac{1}{2\pi i} \int_{\zeta \in \partial G_\epsilon} \frac{a_i(\zeta)d\zeta}{\zeta - z_1} + a_i(z_1)$$

$$= a_i(z_1),$$

where

$$G_\epsilon = \{\zeta \in \Delta(s) \,\big|\, |\operatorname{Im} \zeta| > \epsilon\} .$$

Hence $a_i(z_1)$ can be extended to a holomorphic function $\tilde{a}_i(z_1)$ on $\Delta(r)$. Let

$$\tilde{P}_j(z_1;Z) = Z^\lambda + \sum_{i=0}^{\lambda-1} \tilde{a}_i(z_1)Z^i .$$

Let $\tilde{X}$ be the set of all $(z_1, \ldots, z_n) \in \Delta^n(r)$ such that $\tilde{P}_j(z_1;z_j) = 0$ for $2 \leq j \leq n$. Then $\tilde{X}$ is a subvariety of dimension 1 in $\Delta^n(r)$ containing $X \cap \Delta^n(r)$. Q.E.D.

<u>Remark</u>. In the proof of (2.22) the step of proving (*) can be avoided if we use directly the projection defined by

$$(z_1, \ldots, z_n) \longrightarrow \lambda_1 z_1^2 + \ldots + \lambda_n z_n^2$$

and if we define X as the common zeros of functions corresponding to those in (†) of the proof of (2.3). Then we can conclude (*) of the proof of (2.22) as a consequence of (2.22).

V. Extension of Holomorphic Maps

Suppose B^n is the open unit ball in $\mathbb{C}^n$ and M is a compact Kähler manifold and $f = B^n - \{0\} \longrightarrow M$ is a holomorphic map. We will consider the problem of extending f to a meromorphic map from B^n to M. Extension to a holomorphic map is impossible as is shown by the counter-example where $M = \mathbb{P}_{n-1}$ and f is the restriction of the canonical map $\mathbb{C}^n - \{0\} \to \mathbb{P}_{n-1}$ $(n \geq 2)$. The problem for $n \geq 3$ was solved by Griffiths [6]. Shiffman [20] improved the result to $n \geq 2$ by proving the following lemma. The proof of this lemma given below is new and is very elementary.

(2.23) Lemma. Suppose $\omega = \sqrt{-1} \sum_{i,j=1}^{2} a_{i\bar{j}} dz_i \wedge d\bar{z}_j$ is a closed C^∞ (1,1)-form on $\Delta^2 - \{0\}$ which is nonnegative (i.e. the matrix $(a_{i\bar{j}})$ is semipositive hermitian). Then there exists a C^∞ function u on $\Delta^2 - \{0\}$ such that $\sqrt{-1}\,\partial\bar{\partial} u = \omega$.

Proof. First we prove

$$(*) \qquad \int_{\Delta^2(\epsilon)-\{0\}} a_{1\bar{1}} \sqrt{-1}\, dz_1 \wedge d\bar{z}_1 \wedge \sqrt{-1} dz_2 \wedge d\bar{z}_2 = O(\epsilon^2) .$$

Take a C^∞ function $\varphi = \varphi(z_1)$ on $\mathbb{C}$ with support in Δ such that $\varphi \equiv 1$ on $\Delta(\frac{1}{2})$ and $0 \leq \varphi \leq 1$. Consider the function

$$f(z_2) = \int_{\Delta} \varphi(z_1) a_{1\bar{1}}(z_1, z_2) \sqrt{-1}\, dz_1 \wedge d\bar{z}_1$$

for $z_2 \in \Delta - \{0\}$. Since $d\omega = 0$, one has

$$\begin{cases} \dfrac{\partial a_{1\bar{1}}}{\partial z_2} = \dfrac{\partial a_{2\bar{1}}}{\partial z_1} \\[2ex] \dfrac{\partial a_{1\bar{1}}}{\partial \bar{z}_2} = \dfrac{\partial a_{1\bar{2}}}{\partial \bar{z}_1} \end{cases}$$

on $\Delta^2 - \{0\}$. Hence

$$\frac{\partial f}{\partial z_2} = \int_{\Delta} \varphi(z_1) \frac{\partial a_{2\bar{1}}}{\partial z_1} \sqrt{-1}\, dz_1 \wedge d\bar{z}_1 = -\int_{\Delta} \frac{\partial \varphi}{\partial z_1} a_{2\bar{1}} \sqrt{-1}\, dz_1 \wedge d\bar{z}_1$$

for $z_2 \in \Delta - \{0\}$. Likewise

$$\frac{\partial f}{\partial \bar{z}_2} = -\int_{\Delta} \frac{\partial \varphi}{\partial \bar{z}_1} a_{1\bar{2}} \sqrt{-1}\, dz_1 \wedge d\bar{z}_1$$

for $z_2 \in \Delta - \{0\}$. Since $\frac{\partial \varphi}{\partial z_1}$ and $\frac{\partial \varphi}{\partial \bar{z}_1}$ are both 0 on $\Delta(\frac{1}{2})$, it follows that $\frac{\partial \varphi}{\partial z_1} a_{2\bar{1}}$ and $\frac{\partial \varphi}{\partial \bar{z}_1} a_{1\bar{2}}$ are C^∞ on Δ^2 . Therefore $\frac{\partial f}{\partial z_2}$ and $\frac{\partial f}{\partial \bar{z}_2}$ admit C^∞ extensions to Δ . As a consequence, $f(z_2)$ is uniformly bounded on $\Delta - \{0\}$. For $\epsilon < \frac{1}{2}$,

$$0 \leq \int_{\Delta^2(\epsilon)-\{0\}} a_{1\bar{1}} \sqrt{-1}\, dz_1 \wedge d\bar{z}_1 \wedge \sqrt{-1}\, dz_2 \wedge d\bar{z}_2$$

$$= \int_{\Delta(\epsilon)-\{0\}} \sqrt{-1}\, dz_2 \wedge d\bar{z}_2 \int_{\Delta(\epsilon)} a_{1\bar{1}} \sqrt{-1}\, dz_1 \wedge d\bar{z}_1$$

$$\leq \int_{\Delta(\epsilon)-\{0\}} \sqrt{-1}\, dz_2 \wedge d\bar{z}_2 \;\; f(z_2) \;=\; O(\epsilon^2) ,$$

which is (*). Analogously we have

$$\int_{\Delta^2(\epsilon)-\{0\}} a_{2\bar{2}} \sqrt{-1}\, dz_1 \wedge d\bar{z}_1 \wedge \sqrt{-1}\, dz_2 \wedge d\bar{z}_2 \;=\; O(\epsilon^2) .$$

Since ω is nonnegative, $|a_{1\bar{2}}|^2 \leq a_{1\bar{1}} a_{2\bar{2}}$. Hence

$$0 \leq \int_{\Delta^2(\epsilon)-\{0\}} |a_{1\bar{2}}| \sqrt{-1}\, dz_1 \wedge d\bar{z}_1 \wedge \sqrt{-1}\, dz_2 \wedge d\bar{z}_2$$

$$\leq \left(\int_{\Delta^2(\epsilon)-\{0\}} a_{1\bar{1}} \sqrt{-1}\, dz_1 \wedge d\bar{z}_1 \wedge \sqrt{-1}\, dz_2 \wedge d\bar{z}_2 \right)^{1/2}$$

$$\left(\int_{\Delta^2(\epsilon)-\{0\}} a_{2\bar{2}} \sqrt{-1}\, dz_1 \wedge d\bar{z}_1 \wedge \sqrt{-1}\, dz_2 \wedge d\bar{z}_2 \right)^{1/2}$$

$$= O(\epsilon^2) .$$

Since $-a_{2\bar{1}} = \overline{a_{1\bar{2}}}$, it follows that

$$\int_{\Delta^2(\epsilon)-\{0\}} |a_{2\bar{1}}| \sqrt{-1}\, dz_1 \wedge d\bar{z}_1 \wedge \sqrt{-1}\, dz_2 \wedge d\bar{z}_2 \;=\; O(\epsilon^2) .$$

In particular, $a_{i\bar{j}}$ is locally L^1 on Δ^2 . Hence we can regard ω as a (1,1)-current on Δ^2 .

We are going to prove that $d\omega = 0$ on Δ^2 when ω is regarded as a (1,1)-current on Δ^2. Fix a C^∞ 1-form η on Δ^2 with compact support. Let $\alpha(\epsilon)$ be the supremum of

$$\left|\int_G \omega \wedge \eta \wedge dr\right|$$

as G runs through all open subsets of $B(\epsilon)$, where

$$r = \left(|z_1|^2 + |z_2|^2\right)^{1/2}$$

and

$$B(\epsilon) = \{r < \epsilon\} .$$

Let $\beta(\epsilon)$ be the supremum of

$$\left|\int_G \omega \wedge d\eta\right|$$

as G runs through all open subsets of $B(\epsilon)$. Since

$$\int_{\Delta^2(\epsilon)} |a_{i\bar{j}}| \sqrt{-1}\, dz_1 \wedge d\bar{z}_1 \wedge \sqrt{-1}\, dz_2 \wedge d\bar{z}_2 = O(\epsilon^2) ,$$

one has

$$\begin{cases} \alpha(\epsilon) = O(\epsilon^2) \\ \beta(\epsilon) = O(\epsilon^2) . \end{cases}$$

By Stokes's theorem,

$$\int_{\partial B(\epsilon)} \omega \wedge \eta - \int_{\partial B(r)} \omega \wedge \eta = \int_{B(\epsilon)-B(r)} \omega \wedge d\eta$$

for $0 < r < \frac{\epsilon}{2} < 1$. By integrating with respect to r from $r = \frac{\epsilon}{2}$ to $r = \epsilon$, one obtains

$$\frac{\epsilon}{2}\int_{\partial B(\epsilon)} \omega \wedge \eta = \int_{r=\frac{\epsilon}{2}}^{\epsilon} dr \int_{\partial B(\epsilon)} \omega \wedge \eta$$

$$= \int_{r=\frac{\epsilon}{2}}^{\epsilon} dr \int_{\partial B(r)} \omega \wedge \eta + \int_{r=\frac{\epsilon}{2}}^{\epsilon} dr \int_{B(\epsilon)-B(r)} \omega \wedge d\eta$$

$$= \int_{B(\epsilon)-B(\frac{\epsilon}{2})} \omega \wedge \eta \wedge dr + \int_{r=\frac{\epsilon}{2}}^{\epsilon} dr \int_{B(\epsilon)-B(r)} \omega \wedge d\eta.$$

It follows that

$$\frac{\epsilon}{2}\left|\int_{\partial B(\epsilon)} \omega \wedge \eta \right| \leq \alpha(\epsilon) + \frac{\epsilon}{2}\beta(\epsilon) .$$

Hence

$$\lim_{\epsilon \to 0^+} \int_{\partial B(\epsilon)} \omega \wedge \eta = 0 .$$

By Stokes's theorem,

$$\int_{\Delta^2} \omega \wedge d\eta = \lim_{\epsilon \to 0^+} \int_{\Delta^2 - B(\epsilon)} \omega \wedge d\eta = - \lim_{\epsilon \to 0^+} \int_{\partial B(\epsilon)} \omega \wedge \eta = 0 .$$

It follows that $d\omega = 0$ on Δ^2 .

By Poincaré's **lemma** for currents, there exists a 1-current σ on Δ^2 such that $\omega = d\sigma$. Write

$$\sigma = \sigma_{1,0} + \sigma_{0,1}$$

where $\sigma_{1,0}$ is a $(1,0)$-current and $\sigma_{0,1}$ is a $(0,1)$-current. Since $\omega = d\sigma$ and ω is $(1,1)$, it follows that

$$\begin{cases} \partial \sigma_{1,0} = 0 \\ \bar{\partial} \sigma_{0,1} = 0 . \end{cases}$$

By using the lemma of Dolbeault-Grothendieck (see (2.A.19) of the Appendix of Chapter 2), one obtains distributions v_1 and v_2 on Δ^2 such that

$$\begin{cases} \sigma_{1,0} = \partial v_1 \\ \sigma_{0,1} = \bar{\partial} v_2 . \end{cases}$$

Let $u = \sqrt{-1}\,(v_2 - v_1)$. Then $\omega = \sqrt{-1}\,\partial\bar{\partial} u$. Since ω is C^∞ on $\Delta^2 - \{0\}$, it follows that u is C^∞ on $\Delta^2 - \{0\}$ (see (2.A.20) of the Appendix of Chapter 2). Q.E.D.

(2.24) Theorem. Suppose B^n is the open unit ball in $\mathbb{C}^n$ $(n \geq 2)$ and M is a compact Kähler manifold. Then every holomorphic map $f\colon B^n - \{0\} \longrightarrow M$ extends to a meromorphic map from B^n to M, i.e. the graph G of f extends to a subvariety of $B^n \times M$.

Proof. Let η be the Kähler form of M and let

$$\omega = f^*\eta + \frac{\sqrt{-1}}{2}\,(dz_1 \wedge d\bar{z}_1 + \ldots + dz_n \wedge d\bar{z}_n) .$$

Let $B^n(\epsilon)$ be the open ball in $\mathbb{C}^n$ with center 0 and radius $\epsilon > 0$. Then the volume V_ϵ of

$$G \cap \left((B^n(\epsilon) - \{0\}) \times M\right)$$

is equal to

$$\int_{B^n(\epsilon)-\{0\}} \omega^n .$$

To finish the proof, by Bishop's theorem (2.14) it suffices to prove that V_ϵ is finite for some $\epsilon > 0$.

We are going to prove that there exists a C^∞ function u on

$$\Omega := \Delta^n\left(\frac{1}{\sqrt{n}}\right) - \{0\}$$

such that

$$\sqrt{-1}\,\partial\bar{\partial}u = \omega .$$

The case $n = 2$ follows from (2.23). Now assume $n \geq 3$. Then $H^2(\Omega,\mathbb{C}) = 0$ and $H^1(\Omega,{}_n\mathcal{O}) = 0$ (cf.(3.2)). $\omega = d\sigma$ for some C^∞ 1-form on Ω. Write

$$\sigma = \sigma_{1,0} + \sigma_{0,1} ,$$

where $\sigma_{1,0}$ is (1,0) and $\sigma_{0,1}$ is (0,1). Since $\sigma_{1,0} = 0$ and $\sigma_{0,1} = 0$, it follows from $H^1(\Omega,{}_n\mathcal{O}) = 0$ that there exist C^∞ functions v_1, v_2 on Ω such that $\sigma_{1,0} = \partial v_1$ and $\sigma_{0,1} = \bar{\partial} v_2$. Then

$$u := \sqrt{-1}\,(v_2 - v_1)$$

satisfies

$$\sqrt{-1}\,\partial\bar{\partial}u = \omega$$

on Ω.

Since $\sqrt{-1}\,\partial\bar{\partial} u$ is nonnegative, u is plurisubharmonic on Ω. Define $u(0)$ as the lim sup of $u(z)$ when $z \longrightarrow 0$ in Ω. Then $u(0) < \infty$ and u is plurisubharmonic on $\Delta^n\left(\frac{1}{\sqrt{n}}\right)$, because the sub-mean-value property of u for every 1-disc can be verified by observing that every 1-disc containing 0 is the limit of a sequence of 1-discs none of which contains 0. Let ρ be a nonnegative C^∞ on $\mathbb{C}^n$ with compact support such that

$$\int_{\mathbb{C}^n} \rho(z)\, \sqrt{-1}\, dz_1 \wedge d\bar{z}_1 \wedge \ldots \wedge \sqrt{-1}\, dz_n \wedge d\bar{z}_n = 1 .$$

For $\nu \geq 1$ let

$$u_\nu(z) = \nu^{2n} \int_{\mathbb{C}^n} u_\nu(z+\zeta)\rho(\nu\zeta)\, \sqrt{-1}\, d\zeta_1 \wedge d\bar{\zeta}_1 \wedge \ldots \wedge \sqrt{-1}\, d\zeta_n \wedge d\bar{\zeta}_n .$$

Then, for ν sufficiently large, u_ν is a C^∞ plurisubharmonic function on $\Delta^n\left(\frac{1}{2\sqrt{n}}\right)$ and the derivatives of u_ν approach u uniformly on compact subsets of $\Delta^n\left(\frac{1}{2\sqrt{n}}\right) - \{0\}$ as $\nu \longrightarrow \infty$. For $0 < \epsilon < \frac{1}{2\sqrt{n}}$, it follows from Fatou's lemma and Stokes's theorem that

$$\int_{B^n(\epsilon)-0} (\sqrt{-1}\,\partial\bar{\partial} u)^n \leq \overline{\lim_{\nu \longrightarrow \infty}} \int_{B^n(\epsilon)} (\sqrt{-1}\,\partial\bar{\partial} u_\nu)^n$$

$$= \lim_{\nu \longrightarrow \infty} \int_{\partial B^n(\epsilon)} \sqrt{-1}\,\bar{\partial} u_\nu \wedge (\sqrt{-1}\,\partial\bar{\partial}\, u_\nu)^{n-1}$$

$$= \int_{\partial B^n(\epsilon)} \sqrt{-1}\,\bar{\partial} u \wedge (\sqrt{-1}\,\partial\bar{\partial} u)^{n-1} < \infty .$$

Q.E.D.

Remarks.

(i) In (2.24) the Kähler condition on M cannot be removed. The following is a counter-example. Let M be obtained from $\mathbb{C}^2-\{0\}$ by identifying z with $2^n z$ for $n \in \mathbb{Z}$ and let $f: \mathbb{C}^2-\{0\} \longrightarrow M$ be the quotient map. Then the graph G of f cannot be extended to a subvariety of $\mathbb{C}^2 \times M$, because $\{0\} \times M \subset \overline{G}$ and $\dim\{0\} \times M = \dim G = 2$.

(ii) Generalizations of (2.23) and (2.24) can be found in [8], [25], and [26].

Appendix of Chapter 2

I. Jensen Measures

(2.A.1) Definition. Suppose X is a compact Hausdorff topological space and suppose A is a $\mathbb{C}$-algebra of complex-valued continuous functions on X which is closed under the sup norm topology and which contains 1. Suppose ϕ is an element of the maximum ideal space of A, i.e. $\phi : A \longrightarrow \mathbb{C}$ is a $\mathbb{C}$-algebra homomorphism preserving 1. A Jensen measure μ for A and ϕ is a positive Borel measure on X such that the following Jensen's inequality holds for every $f \in A$:

$$\log|\phi(f)| \leq \int_X \log|f| \, d\mu .$$

Remark. A trivial property of a Jensen measure is that

$$\operatorname{Re} \phi(g) = \int_X \operatorname{Re} g \, d\mu \qquad \text{for} \quad g \in A ,$$

because $e^g, e^{-g} \in A$. In particular,

$$\phi(g) = \int_X g \, d\mu \qquad \text{for} \quad g \in A ,$$

because $\operatorname{Im} g = \operatorname{Re}(-ig)$.

(2.A.2) Theorem. For every A and every ϕ, a Jensen measure always exists.

Proof. Let Q be the set of all real-valued continuous functions u on X such that, for some $c > 0$ and some $f \in A$, with $\phi(f) = 1$, $u > c \log|f|$. Q is open in the space

$C_{\mathbb{R}}(X)$ of all real-valued continuous functions on X with the sup norm topology. Since $|\phi(f)| \leq \|f\|$ for $f \in A$, $0 \notin Q$. Since for $u \in Q$ we can choose a rational $c > 0$ such that $u > c \log|f|$ for some $f \in A$ with $\phi(f) = 1$, it follows that Q is convex. Hence there exists a continuous linear functional ℓ on $C_{\mathbb{R}}(X)$ such that $\ell(u) > 0$ for all $u \in Q$. There exists a finite Borel measure μ on X representing ℓ. Since Q contains all strictly positive continuous functions, ℓ is a positive functional and therefore μ is positive. We can normalize μ so that

$$\int_X d\mu = 1 .$$

We claim that μ is a Jensen measure for A and ϕ. Since

$$\int_X d\mu = 1 ,$$

it suffices to verify Jensen's inequality for $f \in A$ satisfying $\phi(f) = 1$. Since $\log(|f|+\epsilon) \in Q$ for $\epsilon > 0$ and $f \in A$ with $\phi(f) = 1$, it follows that

$$\int_X \log(|f|+\epsilon)d\mu > 0$$

for $\epsilon > 0$ and $f \in A$ with $\phi(f) = 1$. Letting $\epsilon \longrightarrow 0$, we obtain

$$\int_X \log|f| \geq 0$$

for $f \in A$ with $\phi(f) = 1$. Q.E.D.

II. Ranks of Holomorphic Maps

We need the following two well-known facts:

α) If $f: X \longrightarrow Y$ is a holomorphic map of complex spaces, then $\dim_x f^{-1}f(x)$ is an upper semi-continuous function of $x \in X$.

β) (Rank Theorem) Suppose f is a holomorphic map from an open neighborhood U of 0 in $\mathbb{C}^n$ to an open neighborhood W of 0 in $\mathbb{C}^m$ such that $f(0) = 0$. Suppose the rank of the complex Jacobian matrix of f is constantly r on U. Then there exist a biholomorphic map σ from Δ^n onto an open neighborhood of 0 in U and a biholomorphic map τ from Δ^m onto an open neighborhood of 0 in W such that

$$(\tau^{-1}f\sigma)(z_1, \ldots, z_n) = (z_1, \ldots, z_r, 0, \ldots, 0)$$

We introduce the following notation. Suppose X is a complex manifold of dimension n, Y is a complex space, and $f: X \longrightarrow Y$ is a holomorphic map. For $x \in X$ denote by $\operatorname{rank}_x J(f)$ the rank of the complex Jacobian matrix of $\tau f \sigma$ at 0, where σ is a biholomorphic map from an open neighborhood of 0 in $\mathbb{C}^n$ onto an open neighborhood of x in X and τ is a biholomorphic map from an open neighborhood of $f(x)$ in Y onto a complex subspace of an open subset of $\mathbb{C}^N$. Clearly $\operatorname{rank}_x J(f)$ is independent of the choice of σ and τ.

(2.A.3) Definition. Suppose $f: X \longrightarrow Y$ is a holomorphic map of complex spaces and $x \in X$.

(i) If X is of pure dimension d, then define

$$\operatorname{rank}_x f = d - \dim_x f^{-1} f(x) .$$

(ii) If $X = \bigcup_{i \in I} X_i$ is the decomposition of X into pure-dimensional components, then define

$$\operatorname{rank}_x f = \max_{x \in X_i} \operatorname{rank}_x (f | X_i) .$$

(iii) Define $\operatorname{rank} f = \sup_{x \in X} \operatorname{rank}_x f$.

Remark. If x is a regular point of X, $\operatorname{rank}_x f$ and $\operatorname{rank}_x J(f)$ may be different (e.g. $X = Y = \mathbb{C}$, $x = 0$, and $f(z) = z^2$). However, from α) and β) it follows that

$$\operatorname{rank} f = \sup\{\operatorname{rank}_x J(f) \,|\, x \text{ is a regular point of } X\} .$$

(2.A.4) Lemma. *Suppose $f: X \longrightarrow Y$ is a holomorphic map of complex spaces. Let R (respectively S) be the set of all regular (respectively singular) points of X. Suppose $k \geq 0$. Let Z be the set of all $x \in R$ such that $\operatorname{rank}_x J(f) \leq k$. Then $Z \cup S$ is a subvariety of X.*

Proof. Since the problem is local in nature, we can assume that $Y = \mathbb{C}^m$, X is a subvariety of pure dimension d in an open subset G of $\mathbb{C}^n$ whose maximum sheaf of ideals is generated by holomorphic functions $g_1, \ldots, g_\ell$ on G, and f is defined by the restrictions to X of holomorphic functions $f_1, \ldots, f_m$ defined on G. Then it is easy to see that $Z \cup S$ equals

$$\bigcap_{\substack{1 \leq i_1 < \ldots < i_{n-d} \leq \ell \\ 1 \leq j_1 < \ldots < j_{k+1} \leq m}} \left\{ x \in X \mid \operatorname{rank}_x J\left(\Phi_{i_1 \ldots i_{n-d}; j_1 \ldots j_{k+1}}\right) \leq n - d + k + 1 \right\} ,$$

where

$$\Phi_{i_1 \ldots i_{n-d}; j_1 \ldots j_{k+1}} : G \longrightarrow \mathbb{C}^{n-d+k+1}$$

is defined by

$$g_{i_1}, \ldots, g_{i_{n-d}}, f_{j_1}, \ldots, f_{j_{k+1}} .$$

Q.E.D.

(2.A.5) Lemma. *Suppose* $F: X \longrightarrow Y$ *is a holomorphic map of complex spaces and* $k \geq 0$. *Let* $\sigma_k(f)$ *be the set of all* $x \in X$ *such that* $\dim_x f^{-1}f(x) \geq k$. *Then* $\sigma_k(f)$ *is a subvariety of* X .

Proof. Clearly we can assume that X is finite-dimensional. We prove by induction on $\dim X$. Since

$$\sigma_k(f) = \bigcup_{i \in I} \sigma_k(f|X_i)$$

if

$$X = \bigcup_{i \in I} X_i$$

is the decomposition of X into irreducible components, we can assume that X is irreducible. Let S be the set of all singular points of X . Let Z be the set of all regular points x of X such that

$$\operatorname{rank}_x J(f) \leq \dim X-k .$$

By (2.A.4) $Z \cup S$ is a subvariety of X. It is clear that

$$\sigma_k(f) = \sigma_k(f|Z \cup S) .$$

If $Z \cup S = X$, then $X = \sigma_k(f)$. If $Z \cup S \neq X$, then $\dim Z \cup S < \dim X$ and, by induction hypothesis $\sigma_k(f|Z \cup S)$ is a subvariety of $Z \cup S$. Q.E.D.

(2.A.6) Lemma. *Suppose* $f: X \longrightarrow Y$ *is a holomorphic map of complex spaces such that* $\operatorname{rank} f \leq k$. *Then* $f(X) = \bigcup_{i=1}^{\infty} A_i$, *where* A_i *is a subvariety of dimension* $\leq k$ *in an open subset of* Y.

Proof. We can assume that X is finite dimensional. Let $d = \dim X$. We prove by induction on d. We can assume that X is Stein and irreducible. Let $r = \operatorname{rank} f$. Let S be the set of all singular points of X and let Z be the set of all regular points x of X such that $\operatorname{rank}_x J(f) < r$. By (2.A.4) $Z \cup S$ is a proper subvariety of X. There exists a holomorphic function h on X such that $h \not\equiv 0$ on X and $h|Z \cup S \equiv 0$. Let

$$A = \{x \in X | h(x) = 0\} .$$

A is of pure dimension $d-1$ and

$$\operatorname{rank}(f|A) \leq k .$$

By induction hypothesis,

$$f(A) = \bigcup_{i=1}^{\infty} B_i ,$$

where B_i is a subvariety of dimension $\leq k$ in an open subset of Y. By the Rank Theorem (i.e. β)),

$$f(X-A) = \bigcup_{i=1}^{\infty} C_i ,$$

where C_i is a regular subvariety of dimension r in an open subset of Y. Q.E.D.

(2.A.7) Theorem. Suppose $f: X \longrightarrow Y$ is a holomorphic map of complex spaces and $k \geq 0$. Let S be the topological closure of the set of all points $x \in X$ such that $\operatorname{rank}_x f \leq k$. Then S is a subvariety of X and $f(S) = \bigcup_{i=1}^{\infty} A_i$, where A_i is a subvariety of dimension $\leq k$ in an open subset of Y.

Proof. Let X_d be the d-dimensional component of X. Let S_d be the set of all $x \in X_d$ such that

$$\dim_x X_d \cap f^{-1}f(x) \geq d-k .$$

By (2.A.5) S_d is a subvariety of X_d. It is clear that

$$S = \bigcup_{\ell \geq 1} \bigcup_{d_1 < \dots < d_\ell} \left(\overline{S_{d_1} \cap \dots \cap S_{d_\ell} - \bigcup_{d \neq d_1, \dots, d_\ell} X_d} \right).$$

Hence S is a subvariety of X. Since $S \subset \bigcup_d S_d$ and $\operatorname{rank}(f|S_d) \leq k$, $\operatorname{rank}(f|S) \leq k$. The theorem now follows from (2.A.6). Q.E.D.

III. Thick Sets

(2.A.8) Lemma. *Suppose* D *and* G *are respectively open subsets of* $\mathbb{C}^k$ *and* $\mathbb{C}^\ell$. *If* A *is a thick set in* D, *then* $A \times G$ *is a thick set in* $D \times G$.

Proof. Suppose the contrary. Then

$$A \times G \subset \bigcup_{j=1}^{\infty} B_j ,$$

where B_j is a subvariety of codimension ≥ 1 in some open subset of $D \times G$. Let $\pi : D \times G \longrightarrow D$ be the projection onto the first factor. Let S_j be the set of all $x \in B_j$ such that $\text{rank}_x(\pi | B_j) \leq \ell - 1$. By (2.A.7) $\pi(S_j)$ is thin in D. We are going to show that

$$A \subset \bigcup_{j=1}^{\infty} \pi(S_j) .$$

Since each B_j is a countable union of compact subsets,

$$A \times G \subset \bigcup_{i=1}^{\infty} C_i ,$$

where C_i is a compact subset of some $B_{j(i)}$. Fix arbitrarily $t \in A$. Since

$$\{t\} \times G = \bigcup_{i=1}^{\infty} (\{t\} \times G) \cap C_i ,$$

by Baire category theorem there exist $1 \leq i < \infty$ and a nonempty open subset U of G such that $\{t\} \times U \subset C_i$. Since

$$\dim B_{j(i)} \leq k + \ell - 1$$

and $\{t\} \times U$ is of pure dimension ℓ, $\{t\} \times U \subset S_{j(i)}$. So $t \in \pi(S_{j(i)})$. Hence

$$A \subset \bigcup_{j=1}^{\infty} \pi(S_j) ,$$

contradicting that A is thick in D. Q.E.D.

(2.A.9) Proposition. *Suppose* D *is an open subset of* $\mathbb{C}^k$, G *is an open subset of* $\mathbb{C}^{\ell}$, *and* V, W *are subvarieties of* $D \times G$ *such that* V *is irreducible. Suppose* A *is a thick set in* D *and for every* $t \in A$ *some nonempty open subset of* $V \cap (\{t\} \times G)$ *is contained in* W. *Then* $V \subset W$.

Proof. We use the following notation. For any subset E of $D \times G$ and $t \in D$, denote by $E(t)$ the set $E \cap (\{t\} \times G)$.

Let $\pi: D \times G \longrightarrow D$ be the projection onto the first factor. Since $\pi(V)$ contains the thick subset A of D, by (2.A.6) $\operatorname{rank}(\pi|V) = k$. Let $d = \dim V$. By (2.A.4) there exists a subvariety S of dimension $\leq d-1$ in V such that S contains the set of all singular points of V and the set of all regular points x of V satisfying $\operatorname{rank}_x J(\pi|V) < k$.

Let T be the set of all points x of V such that $\operatorname{rank}_x(\pi|V) < k$. Let Y be the set of all points x of S such that $\operatorname{rank}_x(\pi|S) < k$. By (2.A.7) $\pi(T)$ and $\pi(Y)$ are both thin in D. Let B be the set of all $t \in A - \pi(T)$ such that no nonempty open subset of $V(t) - S(t)$ is contained in W. We are going to prove that $B \subset \pi(Y)$. Take $t \in B$.

Since $t \notin \pi(T)$, $V(t)$ has pure dimension $d-k$. Since some open subset of $V(t)$ is contained in W and no nonempty open subset of $V(t)-S(t)$ is contained in W , it follows that $S(t)$ contains a nonempty open subset U of $V(t)$. Since $\dim S \leq d-1$ and U has pure dimension $d-k$, $U \subset Y$. Hence $t \in \pi(Y)$.

Since $\text{rank}_x J(\pi|V-S) = k$ for all $x \in V-S$, we can cover $V-S$ by a countable number of open subsets Ω_i $(1 \leq i < \infty)$ such that for every $1 \leq i < \infty$ there is a commutative diagram

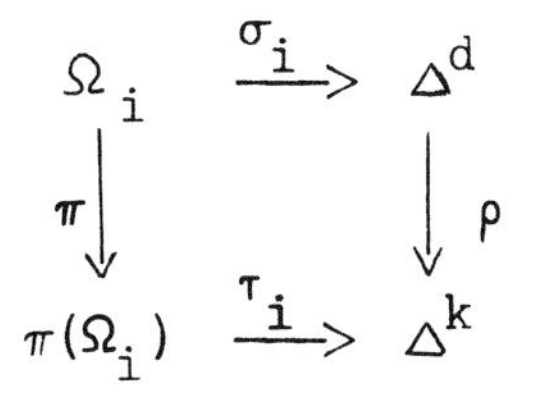

where σ_i and τ_i are biholomorphic maps and

$$\rho(z_1, \ldots, z_d) = (z_1, \ldots, z_k) .$$

Let

$$F_i = \{t \in \pi(\Omega_i) \cap A-\pi(T)-B \,|\, \Omega_i(t) \subset W\} ,$$

$1 \leq i < \infty$. We have

$$A - \pi(T) - B = \bigcup_{i=1}^{\infty} F_i ,$$

because for every $t \in A-\pi(T)-B$ there exists some $1 \leq i < \infty$ such that some nonempty open subset of $\Omega_i(t)$ is contained in W .

For some $1 \leq i < \infty$, F_i is thick in D. By (2.A.8), $\Omega_i \cap \pi^{-1}(F_i)$ is thick in Ω_i. $W \cap \Omega_i$ is the set of all common zeros of a set of holomorphic functions f_j $(j \in J)$ on Ω_i. Since f_j vanishes identically on the thick subset $\Omega_i \cap \pi^{-1}(F_i)$ of Ω_i, f_j vanishes identically on Ω_i $(j \in J)$. Hence $\Omega_i \subset W$. Since V is irreducible, $V \subset W$.

Q.E.D.

IV. Special Analytic Polyhedra

Suppose X is a complex space and A is a set of holomorphic functions on X. We say that X is A-complete if for every $x \in X$ there exists an open neighborhood U of x in X such that $\{x\} = \{y \in U \mid f(x) = f(y) \text{ for all } f \in A\}$.

(2.A.10) Lemma. Suppose $(X,\mathcal{O})$ is a complex space of dimension $\leq d$ and F is a topological vector space with a continuous monomorphism $F \hookrightarrow \Gamma(X,\mathcal{O})$ such that X is F-complete. Let $G = \{g \in F^d \mid X \text{ is g-complete}\}$. Then $F^d - G$ is a countable union of nowhere dense closed subsets of F^d.

Proof. Since X is F-complete, there exist for every $1 \leq i < \infty$ an open subset Ω_i of X, a compact subset K_i of Ω_i, and a holomorphic map

$$\varphi_i : \Omega_i \longrightarrow \mathbb{C}^{n_i}$$

defined by elements of F such that $X = \bigcup_{i=1}^{\infty} K_i$, $\varphi_i(\Omega_i)$ is a subvariety of an open subset of $\mathbb{C}^{n_i}$, and $\varphi_i^{-1}(x)$ has dimension ≤ 0 for every $x \in \mathbb{C}^{n_i}$.

Let E_i be the set of all $(g_1, \ldots, g_d) \in F^d$ such that for some $x \in K_i$ the dimension of the subvariety

$$\{y \in X \mid g_j(y) = g_j(x) \text{ for } 1 \leq j \leq d\}$$

at x is > 0. E_i is a closed subset of F^d and

$$F^d - G = \bigcup_{i=1}^{\infty} E_i .$$

We are going to show that E_i is nowhere dense in F^d. Take $f_1, \ldots, f_d \in F$ and an open neighborhood U of $(f_1, \ldots, f_d)$ in F^d. Let φ_i be defined by $f_{d+1}, \ldots, f_m \in F$ (where $m = n_i+d$). Let $\psi: \Omega \longrightarrow \mathbb{C}^m$ be defined by $f_1, \ldots, f_m$. Then $\psi(\Omega_i)$ is a subvariety of an open subset of $\mathbb{C}^m$. Fix $\epsilon > 0$. By (2.2) there exists a matrix $(a_{\lambda\mu})_{\substack{1 \leq \lambda \leq d \\ 1 \leq \mu \leq m}}$ of complex numbers such that $|a_{\lambda\mu} - \delta_{\lambda\mu}| < \epsilon$ for $1 \leq \lambda \leq d$, $1 \leq \mu \leq m$ (where $\delta_{\lambda\mu}$ is the Kronecker delta) and

$$\dim_x \psi(\Omega_i) \cap (x+T) \leq 0$$

for $x \in \mathbb{C}^m$, where T is the $(m-d)$-plane in $\mathbb{C}^m$ defined by

$$\sum_{\mu=1}^{m} a_{\lambda\mu} z_\mu = 0$$

$(1 \leq \lambda \leq d)$ and $x+T$ is the translate of T by x. Let

$$g_\lambda = \sum_{\mu=1}^{m} a_{\lambda\mu} f_\mu$$

$(1 \leq \lambda \leq d)$. Then

$$(g_1, \ldots, g_d) \in F^d - E_i .$$

When ϵ is sufficiently small,

$$(g_1, \ldots, g_d) \in U .$$

E_i is nowhere dense in F^d. Q.E.D.

(2.A.11) Definition. Suppose X is a complex space and D is an open subset of X .

(a) D is called a polyhedral region if ∂D is compact and there exist an open neighborhood U of ∂D and holomorphic functions $f_1, \ldots, f_k$ on U such that

$$U \cap D = \{x \in U \mid |f_i(x)| < 1 \text{ for } 1 \leq i \leq k\} .$$

$(U, f_1, \ldots, f_k)$ is called a frame for D .

(b) A polyhedral region D with frame $(U, f_1, \ldots, f_k)$ is called unreduced if $k > \dim_x X$ for every $x \in \bar{D}$.

(c) A polyhedral region D with frame $(U, f_1, \ldots, f_k)$ is called prepared if the fibers of the map $U \cap \{f_1 \neq 0\} \longrightarrow \mathbb{C}^{k-1}$ defined by $\frac{f_2}{f_1}, \ldots, \frac{f_k}{f_1}$ are all 0-dimensional.

(d) A polyhedral region D with frame $(U, f_1, \ldots, f_k)$ is called an analytic polyhedron if $D \subset\subset X$ and $\bar{D} \subset U$.

(e) An analytic polyhedron D with frame $(U, f_1, \ldots, f_k)$ is called special if $\dim_x X = k$ for $x \in \bar{D}$.

(2.A.12) Lemma. Suppose $(X,\mathcal{O})$ is a complex space and D is an unreduced polyhedral region in X with frame $(U,f_1, \ldots, f_k)$. Suppose F is a Fréchet space with a continuous monomorphism $F \hookrightarrow \Gamma(U,\mathcal{O})$ such that U is F-complete and $f_1, \ldots, f_k \in F$. Suppose S is a closed subset of X contained in D. Then there exist $g_1, \ldots, g_k \in F$ such that $g_1, \ldots, g_k$ are the functions in a frame for a prepared polyhedral region Q satisfying $S \subset Q \subset D$.

Proof. By shrinking U, we can assume without loss of generality that $\dim U < k$. Let V be a relatively compact open neighborhood of ∂D in $U-S$. Since $\partial V \cap D = \partial V \cap \overline{D}$ is a compact subset of $D \cap U$, there exists $\eta > 0$ such that $\|f_i\|_{\partial V \cap D} \leq 1-\eta$ $(1 \leq i \leq k)$, where $\|\cdot\|_{\partial V \cap D}$ denotes the sup norm on $\partial V \cap D$.

Choose $0 < \epsilon < \frac{\eta}{2}$. Let

$$U' = U \cap \{f_1 \neq 0\}.$$

Let

$$\alpha: \Gamma(U,\mathcal{O}) \longrightarrow \Gamma(U',\mathcal{O})$$

be defined by $\alpha(g) = \frac{g}{f_1}\Big|U'$. Let

$$\beta: F \longrightarrow \Gamma(U',\mathcal{O})$$

be the composite of $F \hookrightarrow \Gamma(U,\mathcal{O})$ and α. Let $F' = F/\operatorname{Ker}\beta$ and let $\overline{\beta}: F' \longrightarrow \Gamma(U',\mathcal{O})$ be induced by β. Since U' is F'-complete, by applying (2.A.10) to the complex space U' and to the continuous monomorphism $\overline{\beta}: F' \longrightarrow \Gamma(U',\mathcal{O})$, we conclude that there exist $h_2, \ldots, h_k \in F$ such that $\|h_i\|_{\overline{V}} < \epsilon$ $(2 \leq i \leq k)$ and the fibers of the map $U' \longrightarrow \mathbb{C}^{k-1}$ defined by

$$\left(\frac{h_2}{f_1} + \frac{f_2}{f_1}, \ldots, \frac{h_k}{f_1} + \frac{f_k}{f_1}\right)$$

are at most 0-dimensional.

Let $g_1 = f_1$ and $g_i = \dfrac{h_i + f_i}{1-\epsilon}$ $(2 \leq i \leq k)$. Let

$$Q = (D-V) \cup \{x \in V \,\big|\, |g_i(x)| < 1 \text{ for } 1 \leq i \leq k\} .$$

We are going to prove that Q is a prepared polyhedral region with frame $(V, g_1, \ldots, g_k)$ and $S \subset Q \subset D$. We have to verify the following four statements:

(i) $S \subset Q$.

(ii) $Q \subset D$.

(iii) Q is an open subset of X.

(iv) $\partial Q \subset V$.

Statement (i) follows from $S \subset D-V$. For (ii) we take $x \in V$ such that $|g_i(x)| < 1$ for $1 \leq i \leq k$. Then

$$|f_1(x)| = |g_1(x)| < 1$$

and, for $2 \leq i \leq k$,

$$\begin{aligned} |f_i(x)| &= |(1-\epsilon)g_i(x) - h_i(x)| \\ &\leq (1-\epsilon)|g_i(x)| + |h_i(x)| \\ &< (1-\epsilon) + \epsilon \\ &= 1 . \end{aligned}$$

Hence $Q \subset D$. To prove (iii), we show first that

$$(*) \qquad \|g_i\|_{\partial V \cap D} < 1 .$$

This follows from

$$\|g_1\|_{\partial V \cap D} = \|f_1\|_{\partial V \cap D} \leq 1-\eta$$

and

$$\|g_i\| \leq \frac{1}{1-\epsilon}\left(\|h_i\|_{\partial V \cap D} + \|f_i\|_{\partial V \cap D}\right)$$

$$\leq \frac{1}{1-\epsilon}(\epsilon+1-\eta) < 1 \qquad (2 \leq i \leq k)$$

It follows from (*) that

$$Q = (D-\overline{V}) \cup \{x \in (U \cap D) \cup V \,\big|\, |g_i(x)| < 1 \text{ for } 1 \leq i \leq k\} .$$

Hence Q is open in X. From the definition of Q it follows that

$$\partial Q \subset \partial(D-V) \cup \overline{V} .$$

Since $Q \subset D$, $\partial Q \subset \overline{D}$. Hence

$$\partial Q \subset V \cup (\partial V \cap D) .$$

Since, by definition of Q, $\partial V \cap D \subset Q$ and since Q is open, (iv) follows. Q.E.D.

(2.A.13) Lemma. *Suppose* X *is a compact metric space and* Y *is a metric space. Suppose* $f\colon X \longrightarrow Y$ *is a continuous map such that the fibers are finite. Then for every* $\epsilon > 0$ *there exists* $\delta > 0$ *such that, if* A *is a set of diameter* $< \delta$ *in*

Y , <u>then every component of</u> $f^{-1}(A)$ <u>has diameter</u> $< \epsilon$.

<u>Proof</u>. Fix $\epsilon > 0$. Suppose the lemma is false. Then there exists, for every $1 \leq n < \infty$, some $A_n \subset Y$ such that the diameter of A_n is $< \frac{1}{n}$ and the diameter of some component B_n of $f^{-1}(A_n)$ is $\geq \epsilon$. Take $a_n, b_n \in B_n$ such that $d(a_n, b_n) \geq \frac{\epsilon}{2}$ (where $d(\cdot,\cdot)$ denotes the distance function). Since X is compact, there exists a subsequence $\{a_{n_k}\}$ of $\{a_n\}$ converging to some $a \in X$. Let

$$f^{-1}f(a) = \{c_1, \ldots, c_\ell\} .$$

Choose $0 < \eta < \frac{\epsilon}{4}$ such that

$$B(c_1,\eta), \ldots, B(c_\ell,\eta)$$

are all mutually disjoint (where $B(c_i,\eta)$ denotes the open ball of radius η centered at c_i). Since X is compact,

$$\bigcup_{i=1}^{\ell} f\big(X - B(c_i,\eta)\big)$$

is closed in Y . Since

$$f(a) \notin \bigcup_{i=1}^{\ell} f\big(X - B(c_i,\eta)\big) ,$$

there exists $\alpha > 0$ such that

$$f^{-1}\big(B(f(a),\alpha)\big) \subset \bigcup_{i=1}^{\ell} B(c_i,\eta) .$$

There exists k_0 such that, for $k \geq k_0$, we have $\frac{1}{n_k} < \frac{\alpha}{2}$ and

$$d\left(f(a_{n_k}),f(a)\right) < \frac{\alpha}{2}.$$

Fix $k \le k_0$. Then $A_{n_k} \subset B(f(a),\alpha)$. Hence $B_{n_k} \subset B(c_{i_k},\eta)$ for some i_k. Since $a_{n_k}, b_{n_k} \in B(c_{i_k},\eta)$, we have

$$d(a_{n_k},b_{n_k}) < 2\eta < \frac{\epsilon}{2},$$

which is a contradiction. Q.E.D.

(2.A.14) Lemma. For $\nu \ge 1$ there exist mutually disjoint open neighborhoods W_p of $e^{\frac{2p\pi i}{\nu}}$ in $\mathbb{C}$ $(1 \le p \le \nu)$ with diameter $\le \frac{4\pi}{\nu}$ such that, if $|a^\nu - 1| < \frac{1}{2}$, then $a \in W_p$ for some $1 \le p \le \nu$.

Proof. Let $a = re^{i\theta}$ where $r > 0$ and $\theta \in \mathbb{R}$. Let $b_\nu = a^\nu - 1$. Suppose $|b_\nu| < \frac{1}{2}$. Then the principal argument of $1+b_\nu$ has absolute value $< \frac{\pi}{2}$. Hence $|\nu\theta - 2n\pi| < \frac{\pi}{2}$ for some integer n. $\left|\theta - \frac{2n\pi}{\nu}\right| < \frac{\pi}{2\nu}$. Let

$$G_{k,\nu} = \left\{ \rho e^{i\phi} \,\middle|\, \rho > 0,\ \phi \in \mathbb{R},\ \left|\phi - \frac{2k\pi}{\nu}\right| < \frac{\pi}{2\nu} \right\}.$$

$G_{k,\nu}$ is an open neighborhood of $e^{\frac{2k\pi i}{\nu}}$ and $G_{1,\nu}, \ldots, G_{\nu,\nu}$ are all mutually disjoint. Since $\left|\theta - \frac{2n\pi}{\nu}\right| < \frac{\pi}{2\nu}$, $a \in G_{k,\nu}$ for some $1 \le k \le \nu$.

We claim that $|1-r| \le \frac{1}{\nu}$. Since

$$(1-|b_\nu|)^{1/\nu} \leq r = |a| \leq (1+|b_\nu|)^{1/\nu},$$

we have $(\frac{1}{2})^{1/\nu} \leq r \leq (\frac{3}{2})^{1/\nu}$. Let $(\frac{3}{2})^{1/\nu} = 1+\alpha_\nu$. Then $\frac{3}{2} \geq 1+\nu\alpha_\nu$. $\alpha_\nu \leq \frac{1}{2\nu}$. Hence $r-1 \leq \frac{1}{2\nu}$. Let $2^{1/\nu} = 1+\beta_\nu$. $2 \geq 1+\nu\beta_\nu$. Hence $\beta_\nu \leq \frac{1}{\nu}$.

$$1 - (\tfrac{1}{2})^{1/\nu} = 1 - \frac{1}{1+\beta_\nu} = \frac{\beta_\nu}{1+\beta_\nu} \leq \beta_\nu < \frac{1}{\nu}.$$

Hence $1-r \leq \frac{1}{\nu}$. The claim is proved.

$$\begin{aligned}
\left|a-e^{\frac{2k\pi i}{\nu}}\right| &= \left(1+r^2-2r\cos(\theta-\tfrac{2k\pi}{\nu})\right)^{1/2} \\
&= \left((1-r)^2+4r\sin^2\tfrac{1}{2}(\theta-\tfrac{2k\pi}{\nu})\right)^{1/2} \\
&\leq |1-r| + 2\sqrt{r}\left|\sin\tfrac{1}{2}(\theta-\tfrac{2k\pi}{\nu})\right| \\
&\leq |1-r| + 2\sqrt{r}\,\frac{\pi}{4\nu} \\
&\leq \frac{1}{\nu} + 2(1+\tfrac{1}{2\nu})\frac{\pi}{4\nu} \\
&< \frac{2\pi}{\nu}.
\end{aligned}$$

Let

$$W_p = G_{p,\nu} \cap \left\{z \in \mathbb{C} \,\middle|\, \left|z-e^{\frac{2k\pi i}{\nu}}\right| < \frac{2\pi}{\nu}\right\},$$

$1 \leq p \leq \nu$. Then W_p satisfies the requirement. Q.E.D.

(2.A.15) Lemma. *Suppose X is a complex space and D is a prepared polyhedral region in X with frame $(U, f_1, \ldots, f_k)$. Suppose $r > 1$ and V is a relatively compact open neighborhood of ∂D in U such that $\|f_i\|_{D \cap \partial V} < \frac{1}{r}$ $(1 \le i \le k)$ (where $\|\cdot\|_{D \cap \partial V}$ denotes the sup norm on $D \cap \partial V$). Let $1 < s < r$ and let R_N be the union of all components of $\{x \in X \mid |(sf_i(x))^N - (sf_1(x))^N| < 1, 2 \le i \le k\}$ which intersect $D-V$. Let*

$$Q_N = \{x \in V \cap R_N \mid |(rf_i(x))^N - (rf_1(x))^N| < 1, 2 \le i \le k\}.$$

Then there exists $N_0 \ge 1$ such that, for $N \ge N_0$, $(D-V) \cup Q_N$ is a polyhedral region with frame

$$\left(V \cap R_N, (rf_2)^N - (rf_1)^N, \ldots, (rf_k)^N - (rf_1)^N\right)$$

and $(D-V) \cup Q_N \subset D$.

Proof. First, we show that, for N sufficiently large, R_N is contained in D. Let T_N be a component of R_N. Suppose $T_N \not\subset D$. Then $T_N \cap \partial D \ne \emptyset$. We will show that there is a contradiction when N is sufficiently large.

Choose $\frac{1}{s} < c < 1$ and let

$$V_0 = \{x \in \bar{V} \mid |f_i(x)| \ge c \text{ for some } 1 \le i \le k\}.$$

Choose $\frac{1}{s} < c_1 < c$ and assume that N is so large that $c_1^N + \frac{1}{s^N} < c^N$. We claim that $|f_1| > c_1$ on $T_N \cap V_0$. Take $x \in T_N \cap V_0$. Then $|f_i(x)| \ge c$ for some $1 \le i \le k$. If $i = 1$, then $|f_1(x)| \ge c > c_1$ and the claim follows. So we can assume that $2 \le i \le k$. Since

$$|(sf_i(x))^N - (sf_1(x))^N| \leq 1 ,$$

it follows that

$$|sf_1(x)|^N \geq (cs)^N - 1$$

and

$$|f_1(x)|^N \geq c^N - \frac{1}{s^N} > c_1{}^N .$$

The claim is proved.

Consider the map

$$\Phi : V_0 \cap \{|f_1| \geq c_1\} \longrightarrow \mathbb{C}^{k-1}$$

defined by $\left(\frac{f_2}{f_1}, \ldots, \frac{f_k}{f_1}\right)$. Now assume that N is so large that $(c_1 s)^N > 2$. For $x \in T_N \cap \bar{D} \cap V_0$ and $2 \leq i \leq k$,

$$\left|\left(\frac{f_i(x)}{f_1(x)}\right)^N - 1\right| = \frac{1}{|sf_1(x)|^N} |(sf_i(x))^N - (sf_1(x))^N| \leq \frac{1}{(c_1 s)^N} < \frac{1}{2} .$$

By (2.A.14), for any component A of $T_N \cap \bar{D} \cap V_0$, the diameter of $\Phi(A)$ is $< \frac{4\pi}{N}\sqrt{k-1}$. Metrize X and let γ be the distance between $D \cap \partial V_0$ and ∂D. Since $V_0 \cap \{|f_1| \geq c_1\}$ is compact and D is prepared, by (2.A.13), when N is sufficiently large, the diameter of every component of $T_N \cap \bar{D} \cap V_0$ is $< \frac{\gamma}{3}$.

Let Γ be the set of all points of $\bar{D} \cap V_0$ whose distances from $D \cap \partial V_0$ and ∂D are $\geq \frac{\gamma}{3}$. Since T_N is connected and intersects both ∂D and $D - V$, there exists $x_0 \in T_N \cap \Gamma$. Let S be the component of $T_N \cap \bar{D} \cap V_0$

containing x_0. Then the diameter of S is $< \frac{\gamma}{3}$. Hence $S \cap (D \cap \partial V_0) = \emptyset$ and $S \cap \partial D = \emptyset$.

Since S is a component of $T_N \cap \bar{D} \cap V_0$, S is closed in $T_N \cap \bar{D} \cap V_0$ and hence is closed in T_N. Since S is disjoint from both ∂D and $D \cap \partial V_0$, S is a component of $T_N \cap D \cap (V_0)^0$ (where $(V_0)^0$ denotes the interior of V_0). Since $T_N \cap D \cap (V_0)^0$ is an open subset of X and is locally connected, S is open in T_N. It follows that $S = T_N$, contradicting $T_N \cap \partial D \neq \emptyset$. Hence R_N is contained in D for N sufficiently large.

Since $\|f_i\|_{D \cap \partial V} < \frac{1}{r}$, for N sufficiently large we have

$$|(rf_i(x))^N - (rf_1(x))^N| < 1$$

on $D \cap \partial V$ $(2 \leq i \leq k)$. Hence $(D-V) \cup Q_N$ is the union of the following two open subsets of X:

$$D - \bar{V}$$

$$\{x \in ((U \cap D) \cup V) \cap R_N \,|\, |(rf_i(x))^N - (rf_1(x))^N| < 1 \text{ for } 2 \leq i \leq k\}.$$

$(D-V) \cup Q_N$ is open in X. To finish the proof, we have only to verify that the boundary B of $(D-V) \cup Q_N$ is a compact subset of $R_N \cap V$. Since $(D-V) \cup Q_N$ is open and $Q_N \subset D$, we have $B \subset V \cap \partial Q_N$. Since ∂Q_N is contained in

$$\{x \in U \cap \bar{R}_N \,|\, |(rf_i(x))^N - (rf_1(x))^N| \leq 1 \text{ for } 2 \leq i \leq k\}$$

and since every $x \in U \cap \partial R_N$ satisfies

$$|(sf_i(x))^N - (sf_1(x))^N| = 1$$

for some $2 \leq i \leq k$, we have $B \subset V \cap R_N$, which implies that B is a compact subset of $V \cap R_N$.   Q.E.D.

(2.A.16) Theorem. Suppose $(X,\mathcal{O})$ is a complex space of dimension $\leq n$ and D is a polyhedral region in X with frame $(U,f_1, \ldots, f_k)$. Suppose F is a Fréchet algebra with a continuous monomorphism $F \hookrightarrow \Gamma(U,\mathcal{O})$ such that U is F-complete and $f_1, \ldots, f_k \in F$. Suppose S is a closed subset of X such that $S \subset D$. Then there exists a polyhedral region Q in X with frame $(V,g_1, \ldots, g_n)$ such that $V \subset U$, $g_1, \ldots, g_n \in F$, and $S \subset Q \subset D$.

Proof. The theorem follows from descending induction on k and (2.A.12) and (2.A.15).   Q.E.D.

(2.A.17) Theorem. Suppose $(X,\mathcal{O})$ is a complex space of pure dimension n and D is an analytic polyhedron in X with frame $(U,f_1, \ldots, f_k)$. Suppose F is a Fréchet algebra with a continuous monomorphism $F \hookrightarrow \Gamma(U,\mathcal{O})$ such that U is F-complete and $f_1, \ldots, f_k \in F$. Suppose S is a closed subset of X such that $S \subset D$. Then there exists a special analytic polyhedron Q in X with frame $(V,g_1, \ldots, g_n)$ such that $V \subset U$, $g_1, \ldots, g_n \in F$, and $S \subset Q \subset D$.

Proof. By (2.A.16) there exists a polyhedral region Q in X with frame $(W,g_1, \ldots, g_n)$ such that $W \subset U$, $g_1, \ldots, g_n \in F$, and $S \subset Q \subset D$. Let G be a relatively compact open neighborhood of ∂Q in W . Since the boundary of the compact set $Q-G$ is $Q \cap \partial G$ and since $|g_i| < 1$ on $Q \cap \partial G$ for $1 \leq i \leq n$, it follows from the maximum modulus principle that $|g_i| < 1$ on $Q-G$ for $1 \leq i \leq n$. Hence Q is a special analytic polyhedron with frame $(Q \cup W, g_1, \ldots, g_n)$. The theorem is satisfied with $V = Q \cup W$.

Q.E.D.

V. A Special Case of the Lemma of Dolbeault-Grothendieck

We present here a simple direct proof of a special case of the lemma of Dolbeault-Grothendieck which is used in (2.23). It is adapted from [29].

First we introduce a notation. Suppose f is a C^∞ function on Δ^2 such that the support of f is contained in $\Delta(r) \times \Delta$ for some $0 < r < 1$. Define

$$(T_1 f)(z_1, z_2) = \frac{1}{2\pi i} \int_{\zeta_1 \in \Delta} \frac{f(\zeta_1, z_2) d\zeta_1 \wedge d\bar{\zeta}_1}{\zeta_1 - z_1} .$$

Then $T_1 f$ is C^∞ on Δ^2 and, since $\operatorname{Supp} f \subset \Delta(r) \times \Delta$, one has

$$\frac{\partial}{\partial \bar{z}_1} T_1 f = T_1 \frac{\partial f}{\partial \bar{z}_1} = f$$

$$\frac{\partial}{\partial z_1} T_1 f = \frac{1}{2\pi i} \int_{\zeta_1 \in \Delta} \frac{\partial f}{\partial \zeta_1} (\zeta_1, z_2) \frac{d\zeta_1 \wedge d\bar{\zeta}_1}{\zeta_1 - z_1}$$

(see [7, pp.24-26]). By using induction on k, from the last equation we conclude that

$$\frac{\partial^k}{\partial z_1{}^k} T_1 f = \frac{1}{2\pi i} \int_{\zeta_1 \in \Delta} \frac{\partial^k f}{\partial \zeta_1{}^k} (\zeta_1, z_2) \frac{d\zeta_1 \wedge d\bar{\zeta}_1}{\zeta_1 - z_1} .$$

Hence, if f_ν is C^∞ on Δ^2 and $\operatorname{Supp} f_\nu \subset \Delta(r) \times \Delta$ and if the derivatives of f_ν converge to the derivatives of f uniformly on compact subsets of Δ^2, then the derivatives of $T_1 f_\nu$ converge to the derivatives of $T_1 f$ uniformly on compact subsets of Δ^2. Analogously we define T_2 and obtain the same conclusion for T_2.

For $0 < r < 1$, let $\mathcal{D}_r$ be the set of all C^∞ functions on Δ^2 with support in $\Delta^2(r)$. $\mathcal{D}_r$ is given the topology of uniform convergence of derivatives on compact supports. Let $\mathcal{D}$ be the set of C^∞ functions on Δ^2 with compact supports. $\mathcal{D}$ is given the topology as the inductive limit of $\{\mathcal{D}_r\}_{r<1}$. Let $\mathcal{D}_{0,q}$ be the set of all C^∞ $(0,q)$-forms on Δ^2 whose coefficients are in $\mathcal{D}$. $\mathcal{D}_{0,q}$ is given the topology as the product of a finite copies of $\mathcal{D}$.

(2.A.18) <u>Lemma</u>. (a) <u>Let E_1 be the closed subsets of all $\omega \in \mathcal{D}_{0,1}$ with $\bar{\partial}\omega = 0$. Then there exists a continuous linear map $\Phi_1 : \mathcal{D}_{0,1} \longrightarrow \mathcal{D}$ such that $\bar{\partial}\Phi_1$ is the identity map of E_1.</u>

(b) <u>Let E_2 be the closed subset of all $\omega \in \mathcal{D}_{0,2}$ such that $\int_{\Delta^2} \omega \wedge f = 0$ for all holomorphic 2-forms on Δ^2. Then there exists a continuous linear map $\Phi_2 : E_2 \longrightarrow \mathcal{D}_{0,1}$ such</u>

that $\bar{\partial}\Phi_2$ is the identity map of E_2 .

Proof. (a) Take $\omega = a_1 d\bar{z}_1 + a_2 d\bar{z}_2 \in E_1$. Let $\rho = \rho(z_2)$ be a C^∞ function on Δ with compact support such that $\rho \equiv 1$ on the support of ω when ρ is regarded naturally as a function on Δ^2 . We have

$$\begin{aligned}\bar{\partial}(\rho T_2 a_2) &= \rho\left(T_2 \frac{\partial a_2}{\partial \bar{z}_1}\right)d\bar{z}_1 + \frac{\partial \rho}{\partial \bar{z}_2}(T_2 a_2)d\bar{z}_2 + \rho a_2 d\bar{z}_2 \\ &= \rho\left(T_2 \frac{\partial a_1}{\partial \bar{z}_2}\right)d\bar{z}_1 + \frac{\partial \rho}{\partial \bar{z}_2}(T_2 a_2)d\bar{z}_2 + \rho\, a_2 d\bar{z}_2 \\ &= \omega + \frac{\partial \rho}{\partial \bar{z}_2}(T_2 a_2)d\bar{z}_2 .\end{aligned}$$

It follows from $\bar{\partial}\omega = 0$ that

$$\frac{\partial}{\partial \bar{z}_1}\left(\frac{\partial \rho}{\partial \bar{z}_2}(T_2 a_2)\right) = 0 .$$

Hence $\frac{\partial \rho}{\partial \bar{z}_2}(T_2 a_2)$ is holomorphic in z_1 . Since $\frac{\partial \rho}{\partial \bar{z}_2}(T_2 a_2)$ has compact support, by the identity theorem of holomorphic functions, $\frac{\partial \rho}{\partial \bar{z}_2}(T_2 a_2)$ is identically zero on Δ^2 . So it suffices to define

$$\Phi_1(\omega) = \rho T_2 a_2 .$$

(b) Take $\omega = a dz_1 \wedge d\bar{z}_2 \in E_2$. The support of ω is contained in $\Delta^2(r)$ for some $0 < r < 1$. Let $b = T_1 T_2 a$. Then

$$\frac{\partial^2 b}{\partial \bar{z}_1 \partial \bar{z}_2} = a .$$

We are going to prove that Supp b is a compact subset of Δ^2.

$$b(z_1,z_2) = \frac{1}{(2\pi i)^2}\int_{\Delta^2} \frac{a(\zeta_1,\zeta_2)d\zeta_1\wedge d\bar{\zeta}_1\wedge d\zeta_2\wedge d\bar{\zeta}_2}{(\zeta_1-z_1)(\zeta_2-z_2)} .$$

Since Supp $a \subset \Delta^2(r)$, it follows that, for fixed $z_1 \in \mathbb{C}$, $b(z_1,\cdot)$ is holomorphic on $\mathbb{C}-\overline{\Delta(r)}$ and, for fixed $z_2 \in \mathbb{C}$, $b(\cdot,z_2)$ is holomorphic on $\mathbb{C}-\overline{\Delta(r)}$. Since

$$\frac{d\zeta_1\wedge d\zeta_2}{(\zeta_1-z_1)(\zeta_2-z_2)}$$

is a holomorphic 2-form on Δ^2 for $(z_1,z_2) \in (\mathbb{C}-\Delta)^2$, it follows from $\omega \in E_2$ that $b(z_1,z_2)$ is 0 on $(\mathbb{C}-\Delta)^2$. From the identity theorem for holomorphic functions we conclude that $b \equiv 0$ on $\mathbb{C}^2-\overline{\Delta^2(r)}$. So Supp b is contained in $\overline{\Delta^2(r)}$ and it suffices to define

$$\Phi_2(\omega) = \frac{\partial b}{\partial\bar{z}_2}d\bar{z}_2 . \qquad \text{Q.E.D.}$$

(2.A.19) <u>Proposition</u>. <u>If</u> σ <u>is a</u> (0,1)-<u>current on</u> Δ^2 <u>such that</u> $\bar\partial\sigma = 0$, <u>then there exists a distribution</u> τ <u>on</u> Δ^2 <u>such that</u> $\bar\partial\tau = \sigma$.

<u>Proof</u>. We use the notations of (2.A.18). Let $\tilde{\sigma} = \sigma\wedge dz_1\wedge dz_2$. Define $\tau' : E_2 \longrightarrow \mathbb{C}$ by

$$\tau'(\omega) = -\tilde{\sigma}(\Phi_2\omega) .$$

By the theorem of Hahn-Banach, τ' can be extended to a continuous linear map $\tilde{\tau}: \mathcal{D}_{0,2} \longrightarrow \mathbb{C}$. Let $\tilde{\tau} = \tau\, dz_1\wedge dz_2$. τ

is a distribution. For $\bar\partial\tau = \sigma$, it suffices to prove $\bar\partial\tilde\tau = \tilde\sigma$.

$$(\bar\partial\tilde\tau)\,\omega = -\tau'(\bar\partial\omega) = \tilde\sigma(\Phi_2\bar\partial\omega)$$

for $\omega \in \mathcal{D}_{0,1}$. Since

$$\bar\partial(\omega - \Phi_2\bar\partial\omega) = \bar\partial\omega - \bar\partial\Phi_2\bar\partial\omega = 0 ,$$

it follows that

$$\omega - \Phi_2\bar\partial\omega = \bar\partial\Phi_1(\omega - \Phi_2\bar\partial\omega) .$$

Hence

$$\tilde\sigma(\omega - \Phi_2\bar\partial\omega) = 0$$

and

$$\tilde\sigma(\Phi_2\bar\partial\omega) = \tilde\sigma(\omega) . \qquad \text{Q.E.D.}$$

(2.A.20) Proposition. If u is a distribution on Δ^2 such that $\partial\bar\partial u$ is a C^∞ (1,1)-form, then u is a C^∞ function.

Proof. Let $\omega = \partial\bar\partial u$. Since $H^2(\Delta^2, \mathbb{C}) = 0$, $\omega = d\sigma$ for some C^∞ 1-form σ on Δ^2. Write $\sigma = \sigma_{1,0} + \sigma_{0,1}$, where $\sigma_{1,0}$ is of type (1,0) and $\sigma_{0,1}$ is of type (0,1). Since $\partial\sigma_{1,0} = 0$ and $\bar\partial\sigma_{0,1} = 0$, it follows from $H^1(\Delta^2, {}_2\mathcal{O}) = 0$ that there exist C^∞ functions v_1, v_2 on Δ^2 such that $\sigma_{1,0} = \partial v_1$ and $\sigma_{0,2} = \bar\partial v_2$. Let $v = v_2 - v_1$. Then v is C^∞ and $\partial\bar\partial v = \omega$. It suffices to show that u-v is C^∞. By replacing u by u-v, we can assume without loss of generality that $\omega = 0$.

Let ρ be a nonnegative C^∞ function on $\mathbb{C}^2$ depending only on $|z_1|^2 + |z_2|^2$ such that

$$\int_{\mathbb{C}^2} \rho(z_1,z_2) \frac{\sqrt{-1}}{2} dz_1 \wedge d\bar{z}_2 \frac{\sqrt{-1}}{2} dz_2 \wedge d\bar{z}_2 = 1 .$$

Let $\rho_\nu = \nu^4 \rho(\nu z)$. Fix arbitrarily $0 < r_1 < r_2 < 1$. For ν sufficiently large, the convolution $u*\rho_\nu$ of u with ρ_ν is C^∞ on $\Delta^2(r_2)$ and satisfies

$$\partial\bar{\partial}(u*\rho_\nu) = 0$$

on $\Delta^2(r_2)$. Hence $u*\rho_\nu$ is a pluriharmonic function on $\Delta^2(r_2)$ and has the mean value property (for ν sufficiently large). It follows that

$$u*\rho_\nu = u*\rho_\nu*\rho_\mu$$

on $\Delta^2(r_1)$ for ν, μ both sufficiently large. By letting $\nu \longrightarrow \infty$, we conclude that $u = u*\rho_\mu$ on $\Delta^2(r_1)$ for μ sufficiently large. Hence u is a C^∞ function. Q.E.D.

CHAPTER 3

HOMOLOGICAL CODIMENSION, LOCAL COHOMOLOGY, AND GAP-SHEAVES

This chapter is devoted to the treatment of the properties of homological codimension, local cohomology, and gap-sheaves which will be needed later on for the extension of coherent subsheaves and sheaves.

(3.1) Lemma. *Suppose* $0 \leq a < b < c$ *in* $\mathbb{R}$ *and* D *is a domain in* $\mathbb{C}^n$. *Let* $G = G^1(a,b)$, $A = \Delta(b)$, *and* $B = G^1(a,c)$. *Suppose* f *is an* L^2 *holomorphic function on* $D \times G$ *and* $f(t,z) = \sum_{k=-\infty}^{\infty} c_k(t) z^k$ *is the Laurent series expansion in* z, *where* $t \in D$ *and* $z \in G$. *Let* $g = \sum_{k=0}^{\infty} c_k z^k$ *on* $D \times A$ *and let* $h = \sum_{k=-\infty}^{-1} c_k z^k$ *on* $D \times B$. *Then there exists* $M > 0$ *independent of* f *such that* $\|g\|_{L^2(D \times A)} \leq M\|f\|_{L^2(D \times G)}$ *and* $\|h\|_{L^2(D \times B)} \leq M\|f\|_{L^2(D \times G)}$ *(where* $\|\cdot\|_{L^2(\Omega)}$ *denotes the* L^2*-norm on* Ω*).*

Proof. Since the general case follows from the special case $n = 0$ by integration over D, we can assume without loss of generality that $n = 0$.

Take $a < a' < b' < b$. Let $A' = \Delta(b')$ and $B' = G^1(a',c)$. Since

$$|g|^2 = |f+h|^2 \leq 2|f|^2 + 2|h|^2$$

on G,

$$\|g\|^2_{L^2(A)} = \|g\|^2_{L^2(A')} + \|g\|^2_{L^2(A-A')}$$

$$\leq \|g\|^2_{L^2(A')} + 2\|f\|^2_{L^2(G)} + 2\|h\|^2_{L^2(B')} < \infty .$$

Likewise,

$$\|h\|^2_{L^2(B)} \leq \|h\|^2_{L^2(B')} + 2\|f\|^2_{L^2(G)} + 2\|h\|^2_{L^2(A')} < \infty .$$

Let H_1 (respectively H) be the Hilbert space of all L^2 holomorphic functions on A (respectively G). Let H_2 be the Hilbert space of all L^2 holomorphic functions h on B such that

$$\int_{|\zeta|=a'} \frac{h(\zeta)d\zeta}{\zeta^k} = 0$$

for all $k \geq 1$. The lemma follows from the bijectivity of the continuous map $H_1 \oplus H_2 \longrightarrow H$ defined by

$$g \oplus h \longrightarrow (g+h)|G . \qquad \text{Q.E.D.}$$

(3.2) Lemma (Frenkel). *Suppose* $0 \leq a < b$ *in* $\mathbb{R}^N$, D *is a Stein domain in* $\mathbb{C}^n$, *and* D' *is a nonempty Stein open subset of* D . *Let* $U_0 = D' \times \Delta^N(b)$, $U_j = D \times (\Delta^N(b) \cap \{|z_j| > a_j\})$ $(1 \leq j \leq N)$, *and* $\mathcal{U} = \{U_i\}_{i=0}^N$. *Then for* $1 \leq \nu \leq N-1$ *there exists a continuous linear map* $\varphi_\nu : Z^\nu(\mathcal{U}, {}_{n+N}\mathcal{O}) \longrightarrow C^{\nu-1}(\mathcal{U}, {}_{n+N}\mathcal{O})$ *such that* $\delta\varphi_\nu =$ *the identity map of* $Z^\nu(\mathcal{U}, {}_{n+N}\mathcal{O})$. *In particular, for* $1 \leq \nu \leq N-1$,

$$H^\nu\big((D' \times \Delta^N(b)) \cup (D \times G^N(a,b)), {}_{n+N}\mathcal{O}\big) = 0 .$$

Moreover, φ_ν *maps* $Z^\nu_{L^2}(\mathfrak{U}, {}_{n+N}\mathcal{O})$ *to* $C^{\nu-1}_{L^2}(\mathfrak{U}, {}_{n+N}\mathcal{O})$ (*where* $C^\nu_{L^2}(\mathfrak{U}, {}_{n+N}\mathcal{O})$ *denotes the set of all*

$$\xi = \{\xi_{i_0 \dots i_\nu}\} \in C^\nu(\mathfrak{U}, {}_{n+N}\mathcal{O})$$

such that $\xi_{i_0 \dots i_\nu}$ is L^2 *on* $U_{i_0} \cap \dots \cap U_{i_\nu}$, *and* $Z^\nu_{L^2}(\mathfrak{U}, {}_{n+N}\mathcal{O})$ *denotes the intersection of* $Z^\nu(\mathfrak{U}, {}_{n+N}\mathcal{O})$ *with* $C^\nu_{L^2}(\mathfrak{U}, {}_{n+N}\mathcal{O})$).

Proof. For a holomorphic function f on $U_{i_0} \cap \dots \cap U_{i_\nu}$ with Laurent series expansion

$$f = \sum_{k=-\infty}^{\infty} c_k z_j^k$$

in z_j, we denote by $e_j(f)$ the function $\sum_{k=0}^{\infty} c_k z_j^k$.

Take

$$\xi = \{\xi_{i_0 \dots i_\nu}\} \in Z^\nu(\mathfrak{U}, {}_{n+N}\mathcal{O})$$

with $1 \leq \nu \leq N-1$. We claim that

$$(1-e_1) \dots (1-e_N)\xi_{i_0 \dots i_\nu} = 0 .$$

Since $(1-e_j)\xi_{i_0 \dots i_\nu} = 0$ whenever $j \notin \{i_0, \dots, i_\nu\}$, it suffices to verify the case

$$\{i_0, \dots, i_\nu\} = \{1, \dots, N\} .$$

For this case we have

$$0 = (\delta\xi)_{01\ldots N} = \xi_{1\ldots N} + \sum_{j=1}^{N} (-1)^j \xi_{01\ldots\hat{j}\ldots N}$$

on $U_0 \cap U_1 \cap \ldots \cap U_N$. Hence

$$(1-e_1) \ldots (1-e_N)\xi_{1\ldots N} = 0$$

on $U_0 \cap U_1 \cap \ldots \cap U_N$. By the identity theorem for holomorphic functions,

$$(1-e_1) \ldots (1-e_N)\xi_{1\ldots N} = 0$$

on $U_1 \cap \ldots \cap U_N$. The claim is proved.

We have

$$(*) \quad \xi_{i_0\ldots i_\nu} = e_1(1-e_2) \ldots (1-e_N)\xi_{i_0\ldots i_\nu} + e_2(1-e_3) \ldots (1-e_N)\xi_{i_0\ldots i_\nu} + \ldots + e_N\xi_{i_0\ldots i_\nu} .$$

Define

$$\varphi_\nu(\xi) = \eta = \{\eta_{i_0\ldots i_{\nu-1}}\} \in C^{\nu-1}(\mathfrak{U}, {}_{n+N}\mathcal{O})$$

by

$$\eta_{i_0\ldots i_{\nu-1}} = (1-e_2) \ldots (1-e_N)\, e_1\, \xi_{1i_0\ldots i_{\nu-1}} + (1-e_3) \ldots (1-e_N)e_2\xi_{2i_0\ldots i_{\nu-1}} + \ldots + (1-e_N)e_{N-1}\xi_{N-1,i_0\ldots i_{\nu-1}} + e_N\xi_{Ni_0\ldots i_{\nu-1}} .$$

Since

$$(\delta\xi)_{ji_0\ldots i_\nu} = \xi_{i_0\ldots i_\nu} + \sum_{k=0}^{\nu}(-1)^{k+1}\xi_{ji_0\ldots\hat{i}_k\ldots i_\nu} = 0$$

we have

$$e_j\xi_{i_0\ldots i_\nu} = \sum_{k=0}^{\nu}(-1)^k e_j\xi_{ji_0\ldots\hat{i}_k\ldots i_\nu}\ .$$

It follows from (*) that $\delta\eta = \xi$. Because of (3.1), φ_ν maps $Z^\nu_{L^2}(\mathcal{U},{}_{n+N}\mathcal{O})$ to $C^{\nu-1}_{L^2}(\mathcal{U},{}_{n+N}\mathcal{O})$. Q.E.D.

Suppose A is a locally closed subset of a topological space X and $\mathcal{F}$ is a sheaf of abelian groups on X . $H^p_A(X,\mathcal{F})$ denotes the pth group of local cohomology of X with supports in A and coefficients in $\mathcal{F}$. Denote by $\mathcal{H}^p_A\mathcal{F}$ the sheaf defined by the presheaf which assigns to every open subset U of X the group $H^p_{A\cap U}(U,\mathcal{F})$ and assigns to the inclusion map of open subsets $V \hookrightarrow U$ the restriction map

$$H^p_{A\cap U}(U,\mathcal{F}) \longrightarrow H^p_{A\cap V}(V,\mathcal{F})\ .$$

We need the following two well-known facts from the theory of local cohomology:

α) If A' is a closed subset of A and $A'' = A-A'$, then one has the following two exact sequences:

$$0 \to H^0_{A'}(X,\mathcal{F}) \to H^0_A(X,\mathcal{F}) \to H^0_{A''}(X,\mathcal{F}) \to H^1_{A'}(X,\mathcal{F}) \to \ldots .$$

$$0 \to \mathcal{H}^0_{A'}\mathcal{F} \to \mathcal{H}^0_A\mathcal{F} \to \mathcal{H}^0_{A''}\mathcal{F} \to \mathcal{H}^1_{A'}\mathcal{F} \to \ldots .$$

β) Suppose $k \geq 0$. If $H^p(X, \mathcal{H}_A{}^q \mathcal{F}) = 0$ for $q \leq k$ and $p \geq 1$, then $H_A{}^p(X, \mathcal{F}) \longrightarrow \Gamma(X, \mathcal{H}_A{}^p \mathcal{F})$ is an isomorphism for $p \leq k+1$.

Suppose M is a finitely generated module over a local ring $(R, \mathfrak{m})$. A sequence $f_1, \ldots, f_k$ in $\mathfrak{m}$ is called an M-sequence if f_j is not a zero-divisor for $M / \sum_{i=1}^{j-1} f_i M$ for $1 \leq j \leq k$. Any permutation of an M-sequence is still an M-sequence. An M-sequence is called maximal if it is not contained in a longer M-sequence. All maximal M-sequences have the same length. This common length is called the homological codimension of M over R, denoted by $\operatorname{codh}_R M$ or simply by $\operatorname{codh} M$. If R is a quotient ring of another local ring S, then it is trivial to see that $\operatorname{codh}_R M = \operatorname{codh}_S M$ when M is reparded naturally as an S-module. If $R = {}_n\mathcal{O}_0$ and if

$$0 \longrightarrow K \longrightarrow R^{p_{\ell-1}} \longrightarrow R^{p_{\ell-2}} \longrightarrow \cdots \longrightarrow R^{p_1} \longrightarrow R^{p_0} \longrightarrow M \longrightarrow 0$$

is exact, then K is free if and only if $\ell + \operatorname{codh} M \geq n$. (All the above statements will be proved in the Appendix of Chapter 3 for the case where R is a quotient ring of some ${}_n\mathcal{O}_0$, which is the only case used in what follows.)

If $\mathcal{F}$ is a coherent analytic sheaf on a complex space $(X, \mathcal{O})$, $\operatorname{codh} \mathcal{F}$ denotes $\min_{x \in X} \operatorname{codh} \mathcal{F}_x$ and $S_m(\mathcal{F})$ denotes

$$\{x \in X \mid \operatorname{codh} \mathcal{F}_x \leq m\}.$$

(3.3) Theorem (Scheja). Suppose X is a complex space, A is a subvariety of X of dimension $\leq d$, and $\mathcal{F}$ is a coherent analytic sheaf on X such that $\operatorname{codh}\mathcal{F} \geq d+q$. Then $\mathcal{H}_A^k\mathcal{F} = 0$ for $0 \leq k < q$, and $H^k(X,\mathcal{F}) \longrightarrow H^k(X-A,\mathcal{F})$ is bijective for $0 \leq k < q-1$ and injective for $k = q-1$.

Proof. Because of statements α) and β) given above, we need only show that $\mathcal{H}_A^k\mathcal{F} = 0$ for $0 \leq k < q$. Since the problem is local in nature, we can assume without loss of generality that X is an open subset of $\mathbb{C}^n$ and we have an exact sequence

$$0 \longrightarrow {}_n\mathcal{O}^{p_{n-d-q}} \longrightarrow \cdots \longrightarrow {}_n\mathcal{O}^{p_1} \longrightarrow {}_n\mathcal{O}^{p_0} \longrightarrow \mathcal{F} \longrightarrow 0$$

on X. Because of this exact sequence, we need only show that $\mathcal{H}_A^k({}_n\mathcal{O}) = 0$ for $0 \leq k < n-d$. If A is regular, this follows from (3.2). If A is not regular, we let A' be the singular set of A and conclude from the exact sequence

$$0 \longrightarrow \mathcal{H}^0_{A'}({}_n\mathcal{O}) \longrightarrow \mathcal{H}_A^0({}_n\mathcal{O}) \longrightarrow \mathcal{H}^0_{A-A'}({}_n\mathcal{O}) \longrightarrow \mathcal{H}^1_{A'}({}_n\mathcal{O}) \longrightarrow \cdots$$

and induction on d that $\mathcal{H}_A^k({}_n\mathcal{O}) = 0$ for $0 \leq k < n-d$.

Q.E.D.

Suppose $\mathcal{F} \subset \mathcal{G}$ are coherent analytic sheaves on a complex space X and A is a subvariety of X. We define the subsheaf $\mathcal{F}[A]_{\mathcal{G}}$ of $\mathcal{G}$ by the following presheaf:

$$U \longmapsto \{s \in \Gamma(U,\mathcal{G}) \,|\, (s|U-A) \in \Gamma(U-A,\mathcal{F})\}.$$

We call $\mathcal{F}[A]_{\mathcal{G}}$ the relative gap-sheaf of $\mathcal{F}$ in $\mathcal{G}$ with respect to A. Note that, when $\mathcal{F} = 0$, $\mathcal{F}[A]_{\mathcal{G}} = \mathcal{H}_A^0\mathcal{G}$.

(3.4) Proposition. Suppose $\mathcal{F} \subset \mathcal{G}$ are coherent analytic sheaves on a complex space X and A is a subvariety of X. Let $\mathcal{I}$ be the ideal-sheaf of A. Then $\mathcal{F}[A]_{\mathcal{G}}$ is coherent and equals $\bigcup_{k=1}^{\infty} (\mathcal{F}:\mathcal{I}^k)_{\mathcal{G}}$ (where the stalk of $(\mathcal{F}:\mathcal{I}^k)_{\mathcal{G}}$ at x is the set of all $s \in \mathcal{G}_x$ such that $\mathcal{I}_x{}^k s \in \mathcal{F}_x$).

Proof. We show first that

$$\mathcal{F}[A]_{\mathcal{G}} = \bigcup_{k=1}^{\infty} (\mathcal{F}:\mathcal{I}^k)_{\mathcal{G}} .$$

Clearly

$$(\mathcal{F}:\mathcal{I}^k)_{\mathcal{G}} \subset \mathcal{F}[A]_{\mathcal{G}} .$$

Suppose U is an open neighborhood of a point x in X and $s \in \Gamma(U, \mathcal{F}[A]_{\mathcal{G}})$. The kernel $\mathcal{J}$ of the sheaf-homomorphism ${}_X\mathcal{O} \longrightarrow \mathcal{G}/\mathcal{F}$ on U defined by multiplication by s is a coherent ideal-sheaf whose zero-set is contained in A. By the Nullstellensatz $\mathcal{I}_x{}^k \subset \mathcal{J}_x$ for some k. Hence $s_x \in \left((\mathcal{F}:\mathcal{I}^k)_{\mathcal{G}}\right)_x$.

Next we show that $(\mathcal{F}:\mathcal{I}^k)_{\mathcal{G}}$ is coherent. Take $x \in X$. There exist an open neighborhood U of x in X and

$$f_1, \ldots, f_{\ell} \in \Gamma(U, \mathcal{I}^k)$$

such that

$$\mathcal{I}^k = \sum_{i=1}^{\ell} f_i \, {}_X\mathcal{O}$$

on U. Then $(\mathcal{F}:\mathcal{I}^k)_{\mathcal{G}} \mid U$ equals the intersection of the kernels of the sheaf-homomorphisms

$$\mathcal{G}|U \longrightarrow (\mathcal{G}/\mathcal{F})|U$$

defined by multiplication by f_i $(1 \leq i \leq \ell)$ and hence is coherent. The coherence of $\mathcal{F}[A]_{\mathcal{G}}$ now follows from the coherence of $\bigcup_{k=1}^{\infty} (\mathcal{F}:\mathcal{J}^k)_{\mathcal{G}}$. Q.E.D.

(3.5) <u>Proposition</u>. <u>If</u> $\mathcal{F}$ <u>is a coherent analytic sheaf on a complex space</u> X , <u>then</u> $S_m(\mathcal{F})$ <u>is a subvariety of dimension</u> $\leq m$.

<u>Proof</u>. We can assume without loss of generality that X is a connected open subset of $\mathbb{C}^n$ and there exists an exact sequence

$$ {}_n\mathcal{O}^{p_{n-m}} \xrightarrow{\alpha} {}_n\mathcal{O}^{p_{n-m-1}} \longrightarrow \cdots \longrightarrow {}_n\mathcal{O}^{p_1} \longrightarrow {}_n\mathcal{O}^{p_0} \longrightarrow \mathcal{F} \longrightarrow 0 $$

on X . α is represented by a matrix (f_{ij}) of holomorphic functions on X . Let

$$ r = \max_{x \in X} \operatorname{rank}\left(f_{ij}(x)\right) . $$

Then $S_m(\mathcal{F})$ equals.

$$ \{x \in X \mid \operatorname{rank}\left(f_{ij}(x)\right) < r\} $$

and hence is a subvariety of X .

To get the dimension estimate of $S_m(\mathcal{F})$, we show first that $S_0(\mathcal{F})$ is 0-dimensional. Fix $x \in X$. Let $\mathcal{G} = \mathcal{F}/0[\{x\}]_{\mathcal{F}}$ (where 0 denotes the zero-subsheaf of $\mathcal{F}$). $\operatorname{codh} \mathcal{G}_x \geq 1$, otherwise there exists $s \in \mathcal{G}_x$ such that $s \neq 0$ and s is annihilated by the maximum ideal of ${}_n\mathcal{O}_x$, contradicting $s \in 0[\{x\}]_{\mathcal{G}} = 0$. There exists an open neighborhood U of x in X such that $U \cap S_0(\mathcal{G}) = \emptyset$. Since

$$S_0(\mathcal{G}) - \{x\} = S_0(\mathcal{F}) - \{x\} ,$$

$\dim S_0(\mathcal{F}) \leq 0$.

For the general case, use induction on m . Suppose $S_m(\mathcal{F})$ has dimension $> m$. At some point $x \in X - S_0(\mathcal{F})$, $S_m(\mathcal{F})$ has dimension $> m$. There exists f_x in the maximum ideal of ${}_n\mathcal{O}_x$ such that f_x is not a zero-divisor for $\mathcal{F}_x$. For some open neighborhood U of x in X , f_x is the germ of some $f \in \Gamma(U, {}_n\mathcal{O})$ such that the sheaf-homomorphism $\mathcal{F} \longrightarrow \mathcal{F}$ on U defined by multiplication by f is injective. We have

$$S_m(\mathcal{F}) \cap U \cap \{f = 0\} \subset S_{m-1}(\mathcal{F}/f\mathcal{F}) .$$

Since by induction hypothesis

$$\dim S_{m-1}(\mathcal{F}/f\mathcal{F}) \leq m-1 ,$$

we have a contradiction. Q.E.D.

Suppose $\mathcal{F} \subset \mathcal{G}$ are coherent analytic sheaves on a complex space X . We define the subsheaf $\mathcal{F}_{[d]\mathcal{G}}$ of $\mathcal{G}$ by the following presheaf:

$$U \longmapsto \{s \in \Gamma(U,\mathcal{G}) \,|\, (s|U-A) \in \Gamma(U-A,\mathcal{F}) \text{ for some subvariety } A \text{ of } U \text{ of dimension} \leq d\} .$$

We call $\mathcal{F}_{[d]\mathcal{G}}$ the dth <u>relative gap-sheaf of</u> $\mathcal{F}$ <u>in</u> $\mathcal{G}$.

(3.6) <u>Theorem.</u> <u>Suppose</u> $\mathcal{F} \subset \mathcal{G}$ <u>are coherent analytic sheaves on a complex space</u> X . <u>Then</u> $\mathcal{F}_{[d]\mathcal{G}} = \mathcal{F}[S_d(\mathcal{F}/\mathcal{G})]_{\mathcal{G}}$. <u>In particular</u>, $\mathcal{F}_{[d]\mathcal{G}}$ <u>is coherent and</u> $\dim \operatorname{Supp}(\mathcal{F}_{[d]\mathcal{G}}/\mathcal{F}) \leq d$.

Proof. The special case where $\mathcal{F} = 0$ follows immediately from (3.3) and (3.5), because, if A is a subvariety of dimension $\leq d$ in an open subset of $X - S_d(\mathcal{G})$, then $\mathcal{H}_A^0 \mathcal{G} = 0$.

Let $\varphi: \mathcal{G} \longrightarrow \mathcal{F}/\mathcal{G}$ be the natural sheaf-epimorphism. The general case follows from the fact that

$$\mathcal{F}_{[d]}\mathcal{G} = \varphi^{-1}\left(0_{[d]}(\mathcal{G}/\mathcal{F})\right) .$$

The last statement of the theorem follows from (3.4) and (3.5). Q.E.D.

(3.7) Lemma. Suppose $\mathcal{F}$ is a coherent analytic sheaf on a complex space $(X, \mathcal{O})$, x is a point of X and $f \in \Gamma(X, \mathcal{O})$. Then f_x is not a zero-divisor for $\mathcal{F}_x$ if and only if $\dim_x V(f) \cap \operatorname{Supp} 0_{[k]}\mathcal{F} < k$ for all $k \geq 0$.

Proof. Suppose f_x is a zero-divisor for $\mathcal{F}_x$. Then there exists $s \in \Gamma(U, \mathcal{F})$ for some open neighborhood U of x in X such that $s_x \neq 0$ and $fs = 0$. Let $k = \dim_x \operatorname{Supp} s$. After shrinking U, we can assume that $\dim \operatorname{Supp} s \leq k$. It follows that $s \in \Gamma(U, 0_{[k]}\mathcal{F})$. Since

$$\operatorname{Supp} s \subset V(f) \cap \operatorname{Supp} 0_{[k]}\mathcal{F},$$

we have

$$\dim_x V(f) \cap \operatorname{Supp} 0_{[k]}\mathcal{F} = k .$$

Conversely suppose

$$\dim_x V(f) \cap \operatorname{Supp} 0_{[k]}\mathcal{F} = k$$

for some $k \geq 0$. Suppose f_x is not a zero-divisor for $\mathcal{F}_x$

and we are going to derive a contradiction. By considering the kernel of the sheaf-homomorphism $\mathcal{F} \longrightarrow \mathcal{F}$ defined by multiplication by f, we conclude that there exists an open neighborhood U of x in X such that, for every $y \in U$, f_y is not a zero-divisor for $\mathcal{F}_y$. Since

$$\dim_x V(f) \cap \operatorname{Supp} 0_{[k]}\mathcal{F} = k ,$$

$V(f)$ contains a k-dimensional branch A of $\operatorname{Supp} 0_{[k]}\mathcal{F}$ which passes through x. For some $y \in A \cap U$ there exists an open neighborhood W of y in U such that

$$A \cap W = W \cap \operatorname{Supp} 0_{[k]}\mathcal{F} .$$

Since

$$W \cap \operatorname{Supp} 0_{[k]}\mathcal{F} \subset V(f) ,$$

after shrinking W we can find a natural number m such that $f^m 0_{[k]}\mathcal{F} | W = 0$. Since $(0_{[k]}\mathcal{F})_y \neq 0$, this contradicts the fact that f_y is not a zero-divisor for $\mathcal{F}_y$. Q.E.D.

(3.8) _Proposition_. _Suppose_ $\mathcal{F}$ _is a coherent analytic sheaf on a complex space_ $(X, \mathcal{O})$ _and_ d _is a nonnegative integer. Then the_ d-_dimensional component of_ $\operatorname{Supp} 0_{[d]}\mathcal{F}$ _equals the_ d-_dimensional component of_ $S_d(\mathcal{F})$. _In other words, for any fixed_ $x \in X$, $\dim_x \operatorname{Supp} 0_{[d]}\mathcal{F} \leq d-1$ _if and only if_ $\dim_x S_d(\mathcal{F}) \leq d-1$.

Proof. We choose to prove the last statement of the proposition. The "if" part follows from the equality

$$\mathcal{O}_{[d]}\mathcal{F} = \mathcal{O}[S_d(\mathcal{F})]_{\mathcal{F}} .$$

For the "only if" part, we assume that

$$\dim_x \operatorname{Supp} \mathcal{O}_{[d]}\mathcal{F} \leq d - 1$$

and $\dim_x S_d(\mathcal{F}) = d$ and we try to derive a contradiction. We can find an open subset U of X such that

$$U \cap \operatorname{Supp} \mathcal{O}_{[d]}\mathcal{F} = \emptyset$$

and

$$\dim U \cap S_d(\mathcal{F}) = d .$$

By replacing U by another open subset, we can assume that there exists $f \in \Gamma(U,\mathcal{O})$ such that the subvariety $V(f)$ defined by f contains $U \cap S_d(\mathcal{F})$ but contains no k-dimensional branch of $U \cap \operatorname{Supp} \mathcal{O}_{[k]}\mathcal{F}$ for any $k > d$. By (3.7) f_y is not a zero-divisor for $\mathcal{F}_y$ for $y \in U$. It follows that

$$S_d(\mathcal{F}) \cap U = S_{d-1}(\mathcal{F}/f\mathcal{F}) \cap U ,$$

contradicting $\dim S_d(\mathcal{F}) \cap U = d$. Q.E.D.

(3.9) <u>Corollary</u>. <u>If</u> $\mathcal{F}$ <u>is a coherent analytic sheaf on a complex space</u> X <u>and</u> d <u>is a nonnegative integer, then</u> $\mathcal{O}_{[d]}\mathcal{F} = 0$ <u>if and only if</u> $\dim S_{k+1}(\mathcal{F}) \leq k$ <u>for</u> $k < d$.

<u>Proof</u>. Follows from (3.8) and the observation that, if

$$\dim \operatorname{Supp} \mathcal{O}_{[k+1]}\mathcal{F} \leq k ,$$

then $\mathcal{O}_{[k+1]}\mathcal{F} = \mathcal{O}_{[k]}\mathcal{F}$. Q.E.D.

(3.10) Corollary. Suppose $\mathcal{F}$ is a coherent analytic sheaf on a complex space $(X,\mathcal{O})$, $x \in X$, and $f \in \Gamma(X,\mathcal{O})$. Let $V(f)$ be the subvariety of X defined by f. Then f_x is not a zero-divisor for $\mathcal{F}_x$ if and only if $\dim_x V(f) \cap S_k(\mathcal{F}) < k$ for all $k \geq 0$.

Proof. Follows from (3.7) and (3.8). Q.E.D.

Suppose G is an open subset of $\mathbb{C}^n$ and $\pi: X \longrightarrow G$ is a holomorphic map of complex spaces and $\mathcal{F}$ is a coherent analytic sheaf on X. $\mathcal{F}$ is said to be π-flat at a point x of X if $t_j - t_j(\pi(x))$ is not a zero-divisor for $\mathcal{F}_x \Big/ \sum_{i=1}^{j-1} \big(t_i - t_i(\pi(x))\big) \mathcal{F}_x$ for $1 \leq j \leq n$, where $t_1, \ldots, t_n$ are the coordinates of $\mathbb{C}^n$. This definition can easily be shown to be equivalent to the usual definition of flatness, but we do not need such an equivalence here.

(3.11) Proposition. Suppose X is a complex space, $\mathcal{F}$ is a coherent analytic sheaf on X, and $\pi: X \longrightarrow \mathbb{C}^n$ is a holomorphic map. Let Z be the set of all $x \in X$ such that $\mathcal{F}_x$ is not π-flat. Then Z is a subvariety of X and rank $\pi|Z < n$.

Proof. By (3.10),

$$Z = \bigcup_{k=0}^{\infty} \{x \in X \,|\, \pi^{-1}\pi(x) \cap S_k(\mathcal{F}) > k-n\} .$$

By (2.A.5) Z is a subvariety of X. Let T_k be the set of all $x \in S_k(\mathcal{F})$ such that

$$\text{rank}_x\ \pi | S_k(\mathcal{F}) < n .$$

Since

$$Z \subset \bigcup_{k=0}^{\infty} T_k ,$$

rank $\pi | Z < n$.   Q.E.D.

Suppose $\mathcal{F}$ is a coherent analytic sheaf on a complex space X . We define the sheaf $\mathcal{F}^{[d]}$ on X by the following presheaf:

$$U \longmapsto \operatorname*{ind.\,lim}_{A \in \mathfrak{A}_d(U)} \Gamma(U-A,\mathcal{F}) ,$$

where $\mathfrak{A}_d(U)$ is the directed set of all subvarieties of U of dimension $\leq d$. We call $\mathcal{F}^{[d]}$ the dth <u>absolute gap-sheaf</u> of $\mathcal{F}$.

(3.12) <u>Proposition</u>. <u>Suppose</u> $\mathcal{F}$ <u>is a coherent analytic sheaf on a complex space</u> $(X,\mathcal{O})$ <u>and</u> d <u>is a nonnegative integer</u>. <u>Then the following three statements are equivalent</u>.

(i) $\mathcal{F}^{[d]}$ <u>is coherent on</u> X .

(ii) $\dim \operatorname{Supp} \mathcal{O}_{[d+1]}\mathcal{F} \leq d$.

(iii) $\dim S_{d+1}(\mathcal{F}) \leq d$.

<u>Proof</u>. The equivalence of (ii) and (iii) follows from (3.8).

(a) (iii) $\Longrightarrow$ (i). We can assume without loss of generality that X is an open subset of $\mathbb{C}^n$ and that there exists an exact sequence

$$(*)\quad 0 \to \mathcal{K} \to {}_n\mathcal{O}^{p_{n-d-3}} \to \cdots$$

$$\to {}_n\mathcal{O}^{p_1} \to {}_n\mathcal{O}^{p_0} \to \mathcal{F} \to 0$$

on X . Let $A = S_{d+1}(\mathcal{F})$ and let

$$i : X - A \hookrightarrow X$$

be the inclusion map. For a sheaf $\mathcal{A}$ on X let $\mathcal{R}_A{}^k\mathcal{A}$ denote the kth direct image of $\mathcal{A}|X-A$ under i . By (3.3)

$$\mathcal{F}^{[d]} = \mathcal{R}_A{}^0\mathcal{F} .$$

Since by (3.3) $\mathcal{R}_A{}^k{}_n\mathcal{O} = 0$ for $1 \leq k < n-d-1$, it follows from the exact sequence (*) that

$$\mathcal{R}_A{}^0\mathcal{F} \approx \mathcal{R}_A{}^{n-d-2}\mathcal{K} .$$

Let

$$\mathcal{K}^* = \mathcal{H}om_{{}_n\mathcal{O}}(\mathcal{K}, {}_n\mathcal{O}) .$$

By replacing X by an arbitrary relatively compact open polydisc in X , we can assume without loss of generality that we have an exact sequence

$$(\dagger)\quad {}_n\mathcal{O}^{q_{n-d-2}} \to \cdots \to {}_n\mathcal{O}^{q_1} \to {}_n\mathcal{O}^{q_0} \to \mathcal{K}^* \to 0$$

on X . Since $\operatorname{codh}\mathcal{F} \geq d+2$ on $X-A$, $\mathcal{K}$ is locally free on $X-A$. By applying the functor $\mathcal{H}om_{{}_n\mathcal{O}}(\cdot, {}_n\mathcal{O})$ to ($\dagger$), we obtain a sequence

$$(\#)\quad 0 \to \mathcal{K} \to {}_n\mathcal{O}^{q_0} \to {}_n\mathcal{O}^{q_1} \to \cdots$$

$$\to {}_n\mathcal{O}^{q_{n-d-3}} \xrightarrow{\alpha} {}_n\mathcal{O}^{q_{n-d-2}}$$

which is exact on $X-A$. Since by (3.3) $\mathcal{R}_A^k {}_n\mathcal{O} = 0$ for $1 \leq k < n-d-1$, it follows from (#) that

$$\mathcal{R}_A^{n-d-2}\mathcal{K} \approx \mathcal{R}_A^0(\operatorname{Im}\alpha) .$$

Since $\mathcal{R}_A^0 {}_n\mathcal{O} = {}_n\mathcal{O}$,

$$\mathcal{R}_A^0(\operatorname{Im}\alpha) \approx (\operatorname{Im}\alpha)[A]_{{}_n\mathcal{O}^{q_{n-d-2}}}$$

which is coherent on X.

(b) (i) $\Longrightarrow$ (ii). Suppose $\dim \operatorname{Supp} 0_{[d+1]}\mathcal{F} = d+1$ and we are going to derive a contradiction. Without loss of generality we can assume that $X = \Delta^n$, $0_{[d]}\mathcal{F} = 0$, and

$$\operatorname{Supp} 0_{[d+1]}\mathcal{F} = \Delta^{d+1} \times \{0\} \subset \Delta^n .$$

Let

$$A = \Delta^d \times \{0\} \subset \Delta^n .$$

Let $\mathcal{G}$ be the 0th direct image of $0_{[d+1]}\mathcal{F}|\Delta^{d+1} \times \{0\}-A$ under the inclusion map $\Delta^{d+1} \times \{0\}-A \hookrightarrow \Delta^{d+1} \times \{0\}$. Since $\mathcal{G} = 0_{[d+1]}\mathcal{F}[A]_{\mathcal{F}[d]}$, $\mathcal{G}$ is coherent. Let f be the $(d+1)^{st}$ coordinate of $\mathbb{C}^n$. Then the sheaf-homomorphism $\mathcal{G} \longrightarrow \mathcal{G}$ defined by multiplication by f is a sheaf-isomorphism. Since f vanishes at 0, by Nakayama's lemma, $\mathcal{G}_0 = 0$. Choose a sequence $\{x_\nu\}$ in $\Delta^{d+1} \times \{0\}-A$ such that $x_\nu \longrightarrow 0$. Since $\Delta^{d+1} \times \{0\}-A$ is Stein, there exists

$$s \in \Gamma(\Delta^{d+1} \times \{0\}-A, 0_{[d+1]}\mathcal{F})$$

such that $s_{x_\nu} \notin \mathfrak{m}_{x_\nu}(0_{[d+1]}\mathcal{F})_{x_\nu}$ for all ν, where $\mathfrak{m}_{x_\nu}$ is the maximum ideal of ${}_n\mathcal{O}_{x_\nu}$. The element s^* of $\mathcal{G}_0$ defined by s is nonzero, contradicting $\mathcal{G}_0 = 0$. Q.E.D.

(3.13) <u>Proposition</u>. <u>Suppose</u> $\mathcal{F}$ <u>is a coherent analytic sheaf on a complex space</u> X <u>and</u> d <u>is a nonnegative integer. Then the natural sheaf-homomorphism</u> $\mathcal{F} \longrightarrow \mathcal{F}^{[d]}$ <u>is a sheaf-isomorphism if and only if</u> $\dim S_{k+2}(\mathcal{F}) \leq k$ <u>for</u> $k < d$.

<u>Proof</u>. (a) The "if" part. Suppose U is an open subset of X and A is a subvariety of dimension $k \leq d$ in U. We are going to prove by induction on k that

$$\Gamma(U,\mathcal{F}) \approx \Gamma(U-A,\mathcal{F}) .$$

Since $\operatorname{codh}\mathcal{F} \geq k+2$ on $U-(A \cap S_{k+1}(\mathcal{F}))$, by (3.3)

$$\Gamma\big(U-(A \cap S_{k+1}(\mathcal{F})), \mathcal{F}\big) \approx \Gamma(U-A,\mathcal{F}) .$$

Since

$$\dim A \cap S_{k+1}(\mathcal{F}) \leq \dim S_{k+1}(\mathcal{F}) < k ,$$

by induction hypothesis

$$\Gamma(U,\mathcal{F}) \approx \Gamma\big(U-(A \cap S_{k+1}(\mathcal{F})), \mathcal{F}\big).$$

It follows that

$$\Gamma(U,\mathcal{F}) \approx \Gamma(U-A,\mathcal{F})$$

and $\mathcal{F} = \mathcal{F}^{[d]}$.

(b) The "only if" part. By (3.12) and the definition of $\mathcal{F}^{[d]}$, we have $0_{[d+1]}\mathcal{F} = 0$. Take arbitrarily $x \in X$. For some open neighborhood U of x in X we can choose $f \in \Gamma(U,{}_X\mathcal{O})$ such that the subvariety $V(f)$ defined by f contains $S_{d+1}(\mathcal{F}) \cap U$ but contains no k-dimensional branch of $\operatorname{Supp} 0_{[k]}\mathcal{F} \cap U$ for any $k > d+1$. By (3.7) f_y is not a zero-divisor for $\mathcal{F}_y$ for $y \in U$. Let $\mathcal{G} = (\mathcal{F}/f\mathcal{F})|U$. Since $\mathcal{F} = \mathcal{F}^{[d]}$, we have $(f\mathcal{F})_{[d]\mathcal{F}} = f\mathcal{F}$ and $0_{[d]\mathcal{G}} = 0$ on U. By (3.9),

$$\dim S_{k+2}(\mathcal{F}) \cap U \leq \dim S_{k+1}(\mathcal{G}) \leq k$$

for $k < d$. Q.E.D.

(3.14) Proposition. *Suppose $0 \leq a < b$ in $\mathbb{R}^N$, D is a domain in $\mathbb{C}^n$, and $\mathcal{F}$ is a coherent analytic sheaf on $D \times \Delta^N(b)$ such that $\mathcal{F}^{[n-1]} = \mathcal{F}$. If D' is a nonempty open subset of D, then the restriction map*

$$\Gamma(D \times \Delta^N(b), \mathcal{F}) \longrightarrow \Gamma\Big((D \times G^N(a,b)) \cap (D' \times \Delta^N(b)), \mathcal{F}\Big)$$

is bijective.

Proof. First observe that, for any open subset U of D, the restriction map

$$\Gamma(U \times \Delta^N(b), \mathcal{F}) \longrightarrow \Gamma(U \times G^N(a,b), \mathcal{F})$$

is injective, because, if $s \in \Gamma(U \times \Delta^N(b), \mathcal{F})$ and $s|U \times G^N(a,b) = 0$, then $\dim \operatorname{Supp} s = 0$ and it follows from $0_{[n]\mathcal{F}} = 0$ that $s = 0$.

Take $s \in \Gamma\Big((D \times G^N(a,b)) \cup (D' \times \Delta^N(b)), \mathcal{F}\Big)$. We use induction on n to prove that s can be extended to an element of $\Gamma(D \times \Delta^N(b), \mathcal{F})$. Consider first the special case where $\operatorname{codh} \mathcal{F} \geq n+1$ on $D \times \Delta^N(b)$. Let Ω be the largest open subset of D such that $s|\Omega \times G^N(a,b)$ extends to an element of $\Gamma(\Omega \times \Delta^N(b), \mathcal{F})$. We have to show that Ω is closed in D. Take a point x of the boundary of Ω in D. Let

$P' \subset P$ be nonempty open polydiscs in D such that $x \in P$ and $P' \subset \Omega$. We have an exact sequence

$$(*) \quad 0 \longrightarrow {}_{n+N}\mathcal{O}^{p_{N-1}} \longrightarrow \dots \longrightarrow {}_{n+N}\mathcal{O}^{p_1} \longrightarrow {}_{n+N}\mathcal{O}^{p_0} \longrightarrow \mathcal{F} \longrightarrow 0$$

on $P \times \Delta^N(b)$ (cf. (5.A.7)). Since by (3.2)

$$H^k\left((P \times G^N(a,b)) \cup (P' \times \Delta^N(b)), {}_{n+N}\mathcal{O}\right) = 0$$

for $1 \leq k \leq N-1$, it follows from (*) that the restriction map

$$\Gamma(P \times \Delta^N(b), \mathcal{F}) \longrightarrow \Gamma\left((P \times G^N(a,b)) \cup (P' \times \Delta^N(b)), \mathcal{F}\right)$$

is bijective. Hence $s|P \times G^N(a,b)$ can be extended to an element of $\Gamma(P \times \Delta^N(b), \mathcal{F})$ and $x \in \Omega$. The special case is proved.

For the general case, take $a < b' < b$ in $\mathbb{R}^N$. Let $\pi: D \times \Delta^N(b) \longrightarrow D$ be the natural projection. Let

$$A = \pi\left(S_n(\mathcal{F}) \cap (D \times \overline{\Delta^N(b')})\right).$$

A is a closed subset of D and is contained in a countable union of local subvarieties of codimension ≥ 2 in D, because $\dim S_n(\mathcal{F}) \leq n-2$ by (3.12). By applying the special case to the domain $D-A$ in $\mathbb{C}^n$, we conclude that $s|(D-A) \times G^N(a,b)$ can be extended to some $s' \in \Gamma((D-A) \times \Delta^N(b), \mathcal{F})$. Take $x \in A$. We are going to show that, for some open neighborhood U of x in D, $s|U \times G^N(a,b)$ can be extended to an element of $\Gamma(U \times \Delta^N(b), \mathcal{F})$. After a linear coordinates transformation in $\mathbb{C}^n$, we can assume without loss of generality that $x = 0$ and, for some $0 < \gamma$ in $\mathbb{R}^{n-2}$ and some $0 < \alpha < \beta$ in $\mathbb{R}^2$, we have

$$\Delta^{n-2}(\gamma) \times \Delta^{2}(\beta) \subset D$$

and

$$(\Delta^{n-2}(\gamma) \times G^{2}(\alpha,\beta)) \cap A = \emptyset .$$

By induction hypothesis, the restriction map

$$\Gamma(\Delta^{n-2}(\gamma) \times \Delta^{2}(\beta) \times \Delta^{N}(b), \mathcal{F}) \longrightarrow$$
$$\Gamma\left((\Delta^{n-2}(\gamma) \times \Delta^{2}(\beta) \times G^{N}(a,b)) \cup (\Delta^{n-2}(\gamma) \times G^{2}(\alpha,\beta) \times \Delta^{N}(b)), \mathcal{F}\right)$$

is surjective. Hence the element of

$$\Gamma\left((\Delta^{n-2}(\gamma) \times \Delta^{2}(\beta) \times G^{N}(a,b)) \cup (\Delta^{n-2}(\gamma) \times G^{2}(\alpha,\beta) \times \Delta^{N}(b)), \mathcal{F}\right)$$

which agrees with s on

$$\Delta^{n-2}(\gamma) \times \Delta^{2}(\beta) \times G^{N}(a,b)$$

and agrees with s' on

$$\Delta^{n-2}(\gamma) \times G^{2}(\alpha,\beta) \times \Delta^{N}(b)$$

can be extended to an element of

$$\Gamma(\Delta^{n-2}(\gamma) \times \Delta^{2}(\beta) \times \Delta^{N}(b), \mathcal{F}) .$$

Q.E.D.

(3.15) Lemma. *Suppose $\mathcal{F}, \mathcal{G}$ are coherent analytic sheaves on a complex space $(X, \mathcal{O})$. If $\mathcal{F} = \mathcal{F}^{[n]}$, then $\mathcal{H}om_{\mathcal{O}}(\mathcal{G},\mathcal{F}) = \mathcal{H}om_{\mathcal{O}}(\mathcal{G},\mathcal{F})^{[n]}$.*

Proof. Since the problem is local in nature, we can assume without loss of generality that there exists an exact sequence

$$\mathcal{O}^q \longrightarrow \mathcal{O}^p \longrightarrow \mathcal{G} \longrightarrow 0$$

on X. We have the following exact sequence

$$0 \longrightarrow \mathcal{H}om_{\mathcal{O}}(\mathcal{G},\mathcal{F}) \longrightarrow \mathcal{H}om_{\mathcal{O}}(\mathcal{O}^p,\mathcal{F}) \longrightarrow \mathcal{H}om_{\mathcal{O}}(\mathcal{O}^q,\mathcal{F})$$

on X. Since

$$\mathcal{H}om_{\mathcal{O}}(\mathcal{O}^k,\mathcal{F}) \approx \underbrace{\mathcal{F} \oplus \ldots \oplus \mathcal{F}}_{k \text{ times}}$$

for $k = p,q$, we have

$$\mathcal{H}om_{\mathcal{O}}(\mathcal{O}^p,\mathcal{F}) = \mathcal{H}om_{\mathcal{O}}(\mathcal{O}^p,\mathcal{F})^{[n]}$$

and $0_{[n]\mathcal{H}om_{\mathcal{O}}(\mathcal{O}^q,\mathcal{F})} = 0$. It follows that

$$\mathcal{H}om_{\mathcal{O}}(\mathcal{G},\mathcal{F}) = \mathcal{H}om_{\mathcal{O}}(\mathcal{G},\mathcal{F})^{[n]} . \qquad \text{Q.E.D.}$$

(3.16) Proposition. *Suppose* $0 \leq a \leq a' < b$ *in* $\mathbb{R}^N$ *and* D *is an open subset of* $\mathbb{C}^n$. *Suppose* $\mathcal{F}_i$ *is a coherent analytic sheaf on* $D \times G^N(a,b)$ *such that* $\mathcal{F}_i = \mathcal{F}_i^{[n]}$ $(i = 1,2)$. *If* $\varphi\colon \mathcal{F}_1 \longrightarrow \mathcal{F}_2$ *is a sheaf-isomorphism on* $D \times G^N(a',b)$, *then* φ *can be uniquely extended to a sheaf-isomorphism* $\mathcal{F}_1 \longrightarrow \mathcal{F}_2$ *on* $D \times G^N(a,b)$.

Proof. It suffices to consider the special case where $a_j = a_j'$ for $2 \leq j \leq N$. By (3.15),

$$\mathcal{H}om_{_{n+N}\mathcal{O}}(\mathcal{F}_1,\mathcal{F}_2) = \mathcal{H}om_{_{n+N}\mathcal{O}}(\mathcal{F}_1,\mathcal{F}_2)^{[n]}$$

and

$$\mathcal{H}om_{_{n+N}\mathcal{O}}(\mathcal{F}_2,\mathcal{F}_1) = \mathcal{H}om_{_{n+N}\mathcal{O}}(\mathcal{F}_2,\mathcal{F}_1)^{[n]} .$$

By (3.14) the following two restriction maps are bijective:

$$\Gamma(D \times G(a,b),\ \mathcal{H}om_{n+N^{\mathcal{O}}}(\mathcal{F}_1,\mathcal{F}_2)) \longrightarrow \Gamma(D \times G(a',b),\ \mathcal{H}om_{n+N^{\mathcal{O}}}(\mathcal{F}_1,\mathcal{F}_2))$$

$$\Gamma(D \times G(a,b),\ \mathcal{H}om_{n+N^{\mathcal{O}}}(\mathcal{F}_2,\mathcal{F}_1)) \longrightarrow \Gamma(D \times G(a',b),\ \mathcal{H}om_{n+N^{\mathcal{O}}}(\mathcal{F}_2,\mathcal{F}_1)).$$

Q.E.D.

(3.17) <u>Proposition</u>. <u>Suppose</u> $k \geq n$, X <u>is a complex space</u>, $\pi: X \to \mathbb{C}^n$ <u>is a holomorphic map</u>, <u>and</u> $\mathcal{F}$ <u>is a coherent analytic sheaf on</u> X <u>with</u> $\mathcal{F}^{[k]} = \mathcal{F}$. <u>Then there exists a thin set</u> A <u>in</u> $\mathbb{C}^n$ <u>such that</u> $\mathcal{F}(t)^{[k-n]} = \mathcal{F}(t)$ <u>for</u> $t \in \mathbb{C}^n - A$, <u>where</u> $\mathcal{F}(t^0) = \mathcal{F} \Big/ \sum_{i=1}^{n} (t_i - t_i^0)\mathcal{F}$ <u>and</u> $t_1,\ldots,t_n$ <u>are the coordinates of</u> $\mathbb{C}^n$.

<u>Proof</u>. Since $\mathcal{F}^{[k]} = \mathcal{F}$, by (3.13) $\dim S_{\ell+2}(\mathcal{F}) \leq \ell$ for $\ell < k$. By (3.11) and (2.A.7) there exists a thin set A in $\mathbb{C}^n$ such that

(i) $\mathcal{F}$ is π-flat on $X - \pi^{-1}(A)$,

(ii) $\dim \pi^{-1}(t) \cap S_{\ell+2}(\mathcal{F}) \leq \ell - n$ for $\ell < k$ and $t \in \mathbb{C}^n - A$.

It follows that, for $t \in \mathbb{C}^n - A$,

$$\dim S_{\ell-n+2}(\mathcal{F}(t)) \leq \ell - n$$

for $\ell < k$. By (3.13), $\mathcal{F}(t)^{[k-n]} = \mathcal{F}(t)$ for $t \in \mathbb{C}^n - A$.

Q.E.D.

(3.18) <u>Proposition</u>. <u>Suppose</u> $k \geq n$, X <u>is a complex space</u>, $\pi: X \longrightarrow \mathbb{C}^n$ <u>is a holomorphic map, and</u> $\mathcal{F}$ <u>is a coherent analytic sheaf on</u> X. <u>Then there exists a thin set</u> A <u>in</u> $\mathbb{C}^n$ <u>such that, for</u> $t \in \mathbb{C}^n - A$, $0_{[k-n]}\mathcal{F}(t) = \left(0_{[k]}\mathcal{F}\right)(t)$.

Proof. Let $\mathcal{G} = 0_{[k]\mathcal{F}}$ and $\mathcal{R} = \mathcal{F}/\mathcal{G}$. Since $0_{[k]\mathcal{R}} = 0$, by (3.9), $\dim S_{\ell+1}(\mathcal{R}) \leq \ell$ for $\ell < k$. By (3.11) and (2.A.7) there exists a thin set A in $\mathbb{C}^n$ such that

(i) $\mathcal{R}$ is π-flat on $X-\pi^{-1}(A)$,

(ii) $\dim \pi^{-1}(t) \cap S_k(\mathcal{F}) \leq k-n$ for $t \in \mathbb{C}^n - A$,

(iii) $\dim \pi^{-1}(t) \cap S_{\ell+1}(\mathcal{R}) \leq \ell-n$ for $\ell < k$ and $t \in \mathbb{C}^n - A$.

Since by (3.6) $\operatorname{Supp} \mathcal{G} \subset S_k(\mathcal{F})$, it follows from (ii) that

(*) $$\dim \operatorname{Supp} \mathcal{G}(t) \leq k-n \quad \text{for } t \in \mathbb{C}^n - A .$$

By (i) and (iii)

$$\dim S_{\ell+1-n}(\mathcal{R}(t)) \leq \ell-n$$

for $\ell < k$ and $t \in \mathbb{C}^n - A$. Hence by (3.9)

(†) $$0_{[k-n]\mathcal{R}(t)} = 0 \quad \text{for } t \in \mathbb{C}^n - A .$$

Since by (i)

$$0 \longrightarrow \mathcal{G}(t) \longrightarrow \mathcal{F}(t) \longrightarrow \mathcal{R}(t) \longrightarrow 0$$

is exact for $t \in \mathbb{C}^n - A$, it follows from (*) and (†) that $\mathcal{G}(t) = 0_{[k-n]\mathcal{F}(t)}$ for $t \in \mathbb{C}^n - A$. Q.E.D.

Appendix of Chapter 3

M-sequences

Suppose $(R,\mathfrak{m})$ is a local ring which is a quotient ring of some ${}_n\mathcal{O}_0$ and suppose M is a finitely generated module over R. We recall two standard facts in commutative algebra.

α) If $P_1, \ldots, P_k$ are the set of all associated prime ideals of the zero-submodule of M, then the set of all zero-divisors of M is $P_1 \cup \ldots \cup P_k$.

β) An ideal P of R is an associated prime ideal of a submodule N of M if and only if there exists $g \in N$ such that $P = \{f \in R | fg \in N\}$. In particular, $\mathfrak{m}$ is an associated prime ideal of N if and only if there exists $g \in M-N$ such that $\mathfrak{m} g \subset N$.

(3.A.1) Proposition. *If $f_1, \ldots, f_r \in \mathfrak{m}$ is an M-sequence, then any permutation of $f_1, \ldots, f_r$ is again an M-sequence.*

Proof. It suffices to prove the case where two adjacent f_i, f_{i+1} are permuted. By replacing M by $M/f_1M + \ldots + f_{i-1}M$, we can assume without loss of generality that $r = 2$.

First we show that f_2 is not a zero-divisor of M. Let

$$N = \{g \in M | f_2 g = 0\} .$$

By Nakayama's lemma, it suffices to prove that $N \subset f_1 N$. Take $g \in N$. Then $f_2 g = 0 \in f_1 M$. Since f_2 is not a zero-divisor of M/f_1M, $g = f_1g'$ for some $g' \in M$. Then $f_2f_1g' = 0$. Since f_1 is not a zero-divisor of M, $f_2g' = 0$. Hence $g' \in N$ and $g = f_1g' \in f_1N$.

Finally we have to prove that f_1 is not a zero-divisor of M/f_2M. Take $g \in M$ such that $f_1g \in f_2M$. We have to show that $g \in f_2M$. Since $f_1g = f_2g'$ for some $g' \in M$ and since f_2 is not a zero-divisor of M/f_1M, it follows that $g' = f_1g''$ for some $g'' \in M$. From $f_1g = f_2f_1g'$ and the fact that f_1 is not a zero-divisor of M, we conclude that $g = f_2g' \in f_2M$. Q.E.D.

(3.A.2) Proposition. All maximal M-sequences have the same length.

Proof. Since R is a quotient ring of some ${}_n\mathcal{O}_0$, we can assume without loss of generality that $R = {}_n\mathcal{O}_0$. We use induction on n. The case $n = 0$ is trivial. Assume $n > 0$. Suppose the proposition is not true. Then we can find a maximal M-sequence $f_1, \ldots, f_r$ of finite length and an M-sequence $g_1, \ldots, g_{r+1}$ (which may not be maximal). Let N be $M/\sum_{i=1}^{r-1} Mf_i$. Let $P_1,\ldots,P_k$ (respectively $Q_1,\ldots,Q_\ell$) be the set of all associated prime ideals of the zero-submodule of N $\left(\text{respectively } M/\sum_{i=1}^{r} g_iM\right)$. Then $\mathfrak{m}$ is not one of $P_1, \ldots, P_k, Q_1, \ldots, Q_\ell$. After a homogeneous linear change of coordinates of $\mathbb{C}^n$, we can assume that the germ of the coordinate function z_n at 0 does not belong to any one of $P_1, \ldots, P_k, Q_1, \ldots, Q_\ell$. Both $f_1, \ldots, f_{r-1}, z_n$ and $g_1, \ldots, g_r, z_n$ are M-sequences.

We are going to prove that $f_1, \ldots, f_{r-1}, z_n$ is a maximal M-sequence. Since f_r forms a maximal N-sequence, $\mathfrak{m}$ is an associated prime ideal of the zero-submodule of N/f_rN. For some $g \in N - f_rN$, we have $\mathfrak{m}g \subset f_rN$. By α), after replacing g by $g + \lambda f_r$ for some integer λ, we can assume that g is not a zero-divisor of N. It follows that $z_n g = f_r h$ for some $h \in N$. We claim that $h \notin z_n N$ and $\mathfrak{m} h \subset z_n N$. (This implies that z_n forms a maximal N-sequence.) Suppose $h = z_n h'$ for some $h' \in N$. Then $z_n g = f_r h = f_r z_n h'$ and $g = f_r h'$, contradicting $g \notin f_r N$. Now take $\alpha \in \mathfrak{m}$. Then $\alpha g = f_r \beta$ for some $\beta \in N$. It follows that

$$\alpha g h = f_r \beta h = z_n \beta g$$

and $\alpha h = z_n \beta \in z_n N$.

Now M/z_nM is a finitely generated module over ${}_{n-1}\mathcal{O}_0$. Let $\bar{f}_i$ (respectively $\bar{g}_j$) be the image of f_i (respectively g_j) in ${}_{n-1}\mathcal{O}_0$. $\bar{f}_1, \ldots, \bar{f}_{r-1}$ is a maximal (M/z_nM)-sequence and $\bar{g}_1, \ldots, \bar{g}_r$ is an (M/z_nM)-sequence. This contradicts the induction hypothesis. Q.E.D.

(3.A.3) Proposition. Suppose $R = {}_n\mathcal{O}_0$. Let $k = k(M)$ be the common length of the maximal M-sequences of M and let

$$0 \longrightarrow K \longrightarrow R^{p_{\ell-1}} \longrightarrow R^{p_{\ell-2}} \longrightarrow \cdots \longrightarrow R^{p_1} \longrightarrow R^{p_0} \longrightarrow M \longrightarrow 0$$

be an exact sequence. Then K is free if and only if $k+\ell \geq n$.

Proof. First we show, by induction on n, that, after a coordinates change of the form

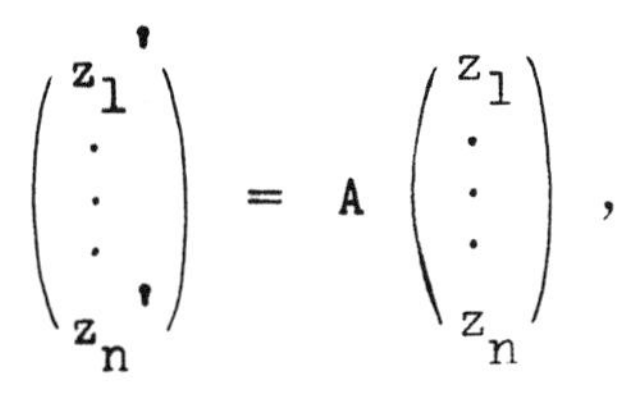

$$\begin{pmatrix} z_1' \\ \cdot \\ \cdot \\ \cdot \\ z_n' \end{pmatrix} = A \begin{pmatrix} z_1 \\ \cdot \\ \cdot \\ \cdot \\ z_n \end{pmatrix},$$

where A is an upper triangular matrix with constant entries, the sequence $z_1, \ldots, z_k$ is an M-sequence. The case $k = 0$ is trivial. Assume $k > 0$. Let $P_1, \ldots, P_q$ be the set of all associated prime ideals of the zero-submodule of M. Then $\mathfrak{m}$ is not any one of $P_1, \ldots, P_q$. After replacing z_1 by

$$z_1' := a_1z_1 + \ldots + a_nz_n \qquad (a_1, \ldots, a_n \in \mathbb{C},\ a_1 \neq 0),$$

we can assume that z_1 does not belong to any one of $P_1, \ldots, P_q$. Then z_1 is not a zero-divisor of M. By applying the induction hypothesis to M/z_1M which can be considered as over ${}_{n-1}\mathcal{O}_0$ (with coordinates $z_2, \ldots, z_n$), we conclude what we intend to show from (3.A.2).

By descending induction on $k(M)$, it is easy to see that the proposition follows from the following two statements:

i) M is free if and only if $k(M) = n$.

ii) If $k(M) < n$ and $0 \longrightarrow N \longrightarrow R^p \longrightarrow M \longrightarrow 0$ is exact, then $k(N) = 1+k(M)$.

The "only if" part of i) is trivial. To prove the "if" part of i), let p be the dimension of the vector space $M/\mathfrak{m}M$ over the field $R/\mathfrak{m}R$. By Nakayama's lemma, M can be generated over R by p elements. There exists an exact sequence

$$0 \longrightarrow N \longrightarrow R^p \longrightarrow M \longrightarrow 0.$$

After a homogeneous linear change of coordinates in $\mathbb{C}^n$, we can assume that $z_1, \ldots, z_n$ form an M-sequence. By induction on q, we conclude that

$$0 \longrightarrow N/z_1N + \ldots + z_qN \longrightarrow (R/z_1R + \ldots + z_qR)^p$$
$$\longrightarrow M/z_1M + \ldots + z_qM \longrightarrow 0$$

is exact. In particular,

$$0 \longrightarrow N/\mathfrak{m}N \longrightarrow (R/\mathfrak{m}R)^p \xrightarrow{\alpha} M/\mathfrak{m}M \longrightarrow 0$$

is exact. Since both vector spaces $(R/\mathfrak{m}R)^p$ and $M/\mathfrak{m}M$ are of dimension p over the field $R/\mathfrak{m}R$, it follows from the surjectivity of α that α is injective. Hence $N/\mathfrak{m}N = 0$. By Nakayama's lemma, $N = 0$. It follows that $M \approx R^p$ is free.

To prove ii), we can assume (after a homogeneous linear change of coordinates in $\mathbb{C}^n$) that $z_1, \ldots, z_k$ form an M-sequence.

$$0 \longrightarrow N/z_1N + \ldots + z_kN \longrightarrow (R/z_1R + \ldots + z_kR)^p$$
$$\longrightarrow M/z_1M + \ldots + z_kM \longrightarrow 0$$

is exact. By replacing M by $M/z_1M + \ldots + z_kM$, we can assume without loss of generality that $k(M) = 0$. We have to show that $k(N) = 1$. Since z_1 is not a zero-divisor of R^p, z_1 is not a zero-divisor of N. It suffices to show that every element $f \in \mathfrak{m}$ is a zero-divisor of N/z_1N. Since f is a zero-divisor of M, there exists $g \in R^p - N$ such that $fg \in N$. Then $z_1fg \in z_1N$ and $z_1g \notin z_1N$. Hence f is a zero-divisor of N/z_1N. Q.E.D.

CHAPTER 4

EXTENSION OF COHERENT ANALYTIC SUBSHEAVES

In this chapter we prove the extension theorem for coherent analytic subsheaves. By using projections involving special analytic polyhedra, we reduce the general case of the extension theorem to a special case where the extension is equivalent to the extendibility of certain meromorphic functions.

(4.1) Lemma. *Suppose* G *is a domain in* $\mathbb{C}^n$, $\mathcal{F}$ *is a coherent analytic sheaf on* G , *and* $\varphi: {}_n\mathcal{O}^r \longrightarrow \mathcal{F}$ *is a sheaf-homomorphism on* G *such that* φ_x *is an isomorphism for some* $x \in G$. *Then* φ *is injective and there exists uniquely a sheaf-homomorphism* $\psi: \mathcal{F} \longrightarrow \mathcal{M}^r$ (*where* $\mathcal{M}$ *is the sheaf of germs of meromorphis functions on* G) *such that*

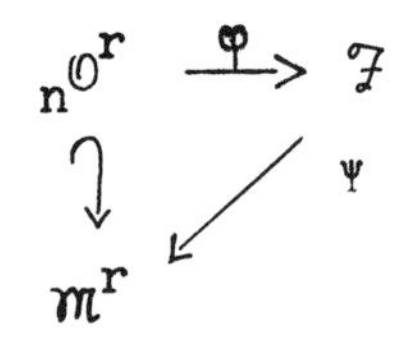

is commutative. Moreover ψ *is injective if* $\mathcal{F}$ *is torsion-free.*

Proof. The injectivity of φ follows from the fact that ${}_n\mathcal{O}^r$ has no torsion and $\operatorname{Supp} \operatorname{Ker} \varphi$ is a proper subvariety of G .

Let $A = \operatorname{Supp} \operatorname{Coker} \varphi$. The uniqueness of ψ follows from the fact that $\mathcal{M}^r$ has no torsion and A is a proper subvariety of G . Because of the uniqueness of ψ , to prove its existence, we need only do it locally. Hence we can assume without loss of generality that there exists a non-identically-zero holomorphic function f on G such that $f\mathcal{F} \subset \operatorname{Im} \varphi$. For $x \in G$ and $s \in \mathcal{F}_x$ define $\psi(s) = f^{-1}\varphi^{-1}(fs)$. ψ satisfies the requirement.

Suppose $\mathcal{F}$ is torsion-free. Then the injectivity of ψ follows from the fact that $\operatorname{Ker} \psi \subset A$. Q.E.D.

(4.2) Lemma. *Suppose $G \subset \tilde{G}$ are open subsets of $\mathbb{C}^n$ and $\mathcal{A}$ is a coherent analytic subsheaf of ${}_n\mathcal{O}^p|G$. Let $\mathcal{F} = {}_n\mathcal{O}^p/\mathcal{A}$ on G and let $s_i \in \Gamma(G,\mathcal{F})$ be the image of $(0, \ldots, 0,1,0, \ldots, 0) \in \Gamma(G,{}_n\mathcal{O}^p)$ under the quotient map ${}_n\mathcal{O}^p \longrightarrow \mathcal{F}$, where the 1 is in the ith position $(1 \leq i \leq p)$. Then $\mathcal{A}$ can be extended coherently to $\tilde{G}$ as a subsheaf of ${}_n\mathcal{O}^p$ if and only if*

(i) *$\mathcal{F}$ can be extended to a coherent analytic sheaf $\tilde{\mathcal{F}}$ on $\tilde{G}$, and*

(ii) *s_i can be extended to some $\tilde{s}_i \in \Gamma(\tilde{G},\tilde{\mathcal{F}})$.*

Proof. For the "if" part, we need only define the extension of $\mathcal{A}$ to be the kernel of the sheaf-homomorphism ${}_n\mathcal{O}^p \longrightarrow \tilde{\mathcal{F}}$ on $\tilde{G}$ defined by $\tilde{s}_1, \ldots, \tilde{s}_p$.

On the other hand, if $\mathcal{A}$ is extended to a coherent analytic subsheaf $\tilde{\mathcal{A}}$ of ${}_n\mathcal{O}^p|G$, then we can define $\tilde{\mathcal{F}}$ to be ${}_n\mathcal{O}^p/\tilde{\mathcal{A}}$ on $\tilde{G}$ and define $\tilde{s}_i \in \Gamma(\tilde{G},\tilde{\mathcal{F}})$ to be the image of $(0, \ldots, 0,1,0, \ldots, 0) \in \Gamma(\tilde{G},{}_n\mathcal{O}^p)$ under the quotient map ${}_n\mathcal{O}^p \longrightarrow \tilde{\mathcal{F}}$, where the 1 is in the ith position. Q.E.D.

Suppose G is a domain in $\mathbb{C}^n$ and $\mathscr{A}$ is a coherent analytic subsheaf of ${}_n\mathcal{O}^p|G$. Let $\mathcal{F} = {}_n\mathcal{O}^p/\mathscr{A}$ on G. Fix $x \in G\text{-}S_{n-1}(\mathcal{F})$. For some open neighborhood U of x in $G\text{-}S_{n-1}(\mathcal{F})$, $\mathcal{F}|U \approx {}_n\mathcal{O}^r|U$ for some r. Let $s_i \in \Gamma(G,\mathcal{F})$ be the image of

$$(0, \ldots, 0,1,0, \ldots, 0) \in \Gamma(G,{}_n\mathcal{O}^p)$$

under the quotient map ${}_n\mathcal{O}^p \longrightarrow \mathcal{F}$, where the 1 is in the ith position $(1 \leq i \leq p)$. We can find $1 \leq i_1 < \ldots < i_r \leq p$ such that $s_{i_1}, \ldots, s_{i_r}$ generates $\mathcal{F}_x$. Let $\varphi\colon {}_n\mathcal{O}^r \longrightarrow \mathcal{F}$ be defined by $s_{i_1}, \ldots, s_{i_r}$. Then φ_x is an isomorphism. By (4.1) there exists a unique sheaf-homomorphism $\psi\colon \mathcal{F} \longrightarrow \mathcal{M}^r$ (where $\mathcal{M}$ is the sheaf of germs of meromorphic functions on G) such that

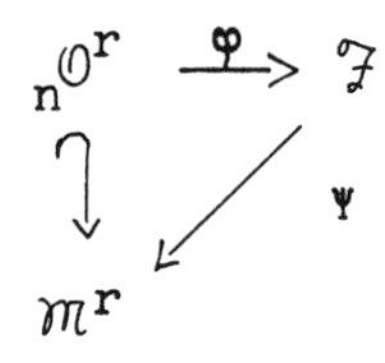

is commutative. Let $t_i = \psi(s_i)$ $(1 \leq i \leq p)$. We call $t_1, \ldots, t_p$ a set of associated meromorphic vector-functions for $\mathscr{A}$.

(4.3) Proposition. Suppose $\tilde{G}$ is a domain in $\mathbb{C}^n$ containing G and suppose $\mathscr{A}_{[n-1]_n\mathcal{O}^p} = \mathscr{A}$. Then $\mathscr{A}$ can be extended to $\tilde{G}$ as a subsheaf of $\tilde{G}$ if and only if $t_1, \ldots, t_p$ can be extended to meromorphic vector-functions on $\tilde{G}$.

Proof. The "only if" part is clear. To prove the "if" part, let $\tilde{t}_i$ be the meromorphic vector-function on $\tilde{G}$ extending t_i $(1 \leq i \leq p)$. Let $\tilde{\mathcal{F}}$ be the analytic subsheaf of $\mathcal{M}^r|\tilde{G}$ generated by $\tilde{t}_1, \ldots, \tilde{t}_p$. $\tilde{\mathcal{F}}$ is coherent, because, if U is a connected open subset of $\tilde{G}$ and f is a non-identically-zero holomorphic function on U such that $f\tilde{t}_i|U$ is a holomorphic vector-function on U $(1 \leq i \leq p)$, then $\tilde{\mathcal{F}}|U$ is isomorphic to

$$\sum_{i=1}^{p} {}_n\mathcal{O} f\tilde{t}_i|U \subset {}_n\mathcal{O}^r|U .$$

Since

$$\mathcal{A}_{[n-1]_n\mathcal{O}^p} = \mathcal{A} ,$$

$\mathcal{F}$ is torsion-free. Hence ψ is injective. If we identify $\mathcal{F}$ with $\psi(\mathcal{F})$, then $\tilde{\mathcal{F}}$ extends $\mathcal{F}$ and $\tilde{t}_i$ extends s_i $(1 \leq i \leq p)$. By (4.2), $\mathcal{A}$ can be extended coherently to $\tilde{G}$ as a subsheaf of ${}_n\mathcal{O}^p$. Q.E.D.

Remark. For the "only if" part of (4.3), it is not necessary to assume $\mathcal{A}_{[n-1]_n\mathcal{O}^p} = \mathcal{A}$.

(4.4) *Lemma. Suppose* $0 \leq a \leq a' < b$ *in* $\mathbb{R}^N$, D *is an open subset of* $\mathbb{C}^n$, $\mathcal{G}$ *is a coherent analytic sheaf on* $D \times \Delta^N(b)$, *and* $\mathcal{F}_i$ *is a coherent analytic subsheaf of* $\mathcal{G}$ *on* $D \times \Delta^N(b)$ *or* $D \times G^N(a,b)$ *such that* $(\mathcal{F}_i)_{[n]\mathcal{G}} = \mathcal{F}_i$ $(i = 1,2)$. *If* $\mathcal{F}_1 = \mathcal{F}_2$ *on* $D \times G^N(a',b)$, *then* $\mathcal{F}_1 = \mathcal{F}_2$ *on their domain of definition.*

Proof. Let V be the set of points where $\mathcal{F}_1$ and $\mathcal{F}_2$ disagree. Since

$$V = \left(\operatorname{Supp} \mathcal{F}_1/(\mathcal{F}_1 \cap \mathcal{F}_2)\right) \cup \left(\operatorname{Supp} \mathcal{F}_2/(\mathcal{F}_1 \cap \mathcal{F}_2)\right) ,$$

V is a subvariety. Since $(\mathcal{F}_i)_{[n]\mathcal{G}} = \mathcal{F}_i$ $(i = 1,2)$, every nonempty branch of V has dimension $> n$. By (2.17)(a), $\dim V \leq n$. Hence $V = \emptyset$. Q.E.D.

Recall the following notation. If $\pi : X \longrightarrow \mathbb{C}^n$ is a holomorphic map and $\mathcal{F}$ is a coherent analytic sheaf on X, then, for $t^0 \in \mathbb{C}^n$, $\mathcal{F}(t^0)$ denotes $\mathcal{F} \Big/ \sum_{i=1}^{n} (t_i - t_i^0)\mathcal{F}$, where $t_1, \ldots, t_n$ are the coordinates of $\mathbb{C}^n$.

(4.5) <u>Theorem</u> (Subsheaf Extension). *Suppose* $0 \leq a < b$ *in* $\mathbb{R}^N$, D *is a domain in* $\mathbb{C}^n$, *and* A *is a thick set in* D. *Suppose* $\mathcal{G}$ *is a coherent analytic sheaf on* $D \times \Delta^N(b)$ *and* $\mathcal{F}$ *is a coherent analytic subsheaf of* $\mathcal{G}$ *on* $D \times G^N(a,b)$ *such that* $\mathcal{F}_{[n]\mathcal{G}} = \mathcal{F}$. *If, for every* $t \in A$, $\operatorname{Im}(\mathcal{F}(t) \longrightarrow \mathcal{G}(t))$ *can be extended coherently to* $\{t\} \times \Delta^N(b)$ *as a subsheaf of* $\mathcal{G}(t)$, *then* $\mathcal{F}$ *can be extended uniquely to a coherent analytic subsheaf* $\tilde{\mathcal{F}}$ *of* $\mathcal{G}$ *on* $D \times \Delta^N(b)$ *such that* $\tilde{\mathcal{F}}_{[n]\mathcal{G}} = \tilde{\mathcal{F}}$.

Proof. The uniqueness of $\tilde{\mathcal{F}}$ follows from (4.4). Moreover, by (4.4), to prove the existence of $\tilde{\mathcal{F}}$, we need only show that, for every $t \in D$, there exists an open neighborhood U of t in D and $a \leq a' < b' \leq b$ in $\mathbb{R}^N$ such that $\mathcal{F}|U \times G^N(a',b')$ can be extended to a coherent analytic subsheaf of $\mathcal{G}|U \times \Delta^N(b')$.

Let k be the largest integer such that $\mathcal{F}_{[k]\mathcal{G}} = \mathcal{F}$. We are going to prove the existence of $\tilde{\mathcal{F}}$ by descending induction on k. When $k \geq n+N$, $\mathcal{F} = \mathcal{G}$ on $D \times G^N(a,b)$ and we can set $\tilde{\mathcal{F}} = \mathcal{G}$. Assume $k < n+N$. By induction hypothesis $\mathcal{F}_{[k+1]\mathcal{G}}$ can be extended to a coherent analytic subsheaf $\mathcal{F}^{\#}$ of $\mathcal{G}$ on $D \times \Delta^N(b)$ (because $k+1 \geq n+1$ and we can use as the thick set the open subset $D \times G^1(a_1,b_1)$ of $D \times \Delta(b_1)$). By replacing $\mathcal{G}$ by $\mathcal{F}^{\#}$, we can assume without loss of generality that $\mathcal{F}_{[k+1]\mathcal{G}} = \mathcal{G}$. Let

$$V = \operatorname{Supp}(\mathcal{G}/\mathcal{F}) \mid D \times G^N(a,b) .$$

Since $\mathcal{F}_{[k]\mathcal{G}} = \mathcal{F}$ and $\mathcal{F}_{[k+1]\mathcal{G}} = \mathcal{G}$, every nonempty branch of V has dimension $k+1$. For a subset E of $D \times \Delta^N(b)$ denote by $E(t)$ the set $E \cap (\{t\} \times \Delta^N(b))$. By Nakayama's lemma,

$$V(t) = \operatorname{Supp}\left(\mathcal{G}(t)/\operatorname{Im}(\mathcal{F}(t) \longrightarrow \mathcal{G}(t))\right) .$$

By (2.18), V can be extended to a subvariety $\tilde{V}$ of $D \times \Delta^N(b)$ whose every nonempty branch has dimension $k+1$. Let $\mathcal{J}$ be the ideal-sheaf of $\tilde{V}$. After shrinking D and $G^N(a,b)$, we can assume without loss of generality that there exists a nonnegative integer m such that $\mathcal{J}^m(\mathcal{G}/\mathcal{F}) = 0$. By replacing $\mathcal{G}$ by $\mathcal{G}/\mathcal{J}^m\mathcal{G}$ and $\mathcal{F}$ by $\mathcal{F}/\mathcal{J}^m\mathcal{F}$, we can assume that $\operatorname{Supp}\mathcal{G} = \tilde{V}$.

(a) Consider first the special case where $\dim \tilde{V}(t) \leq k+1-n$ for $t \in A$. Let A' be the set of all $t \in D$ such that, for every open neighborhood U of t in D, $A \cap U$ is thick in U. A' is thick in D. We are going to show that for every $t \in A'$ there exists an open neighborhood U of t in D such that $\mathcal{F} \mid U \times G^N(a,b)$ can be extended to a coherent analytic subsheaf of $\mathcal{G} | U \times \Delta^N(b)$.

Fix $t^0 \in A$. If $\tilde{V}(t^0) = \emptyset$. then for $a < b' < b$ in $\mathbb{R}^N$ there exists an open neighborhood U of t^0 in D such that $\tilde{V} \cap (U \times \overline{\Delta^N(b')}) = \emptyset$ and $\mathcal{F} = \mathcal{G}$ on $U \times G^N(a,b')$. So we can assume that $\tilde{V}(t^0)$ has pure dimension $k+1-n$.

Let $\ell = k+1-n$. Take $a < a' < b' < b$ in $\mathbb{R}^N$. By applying the theorem on the existence of special analytic polyhedra (i.e. (2.A.17)) to $\tilde{V}(t^0)$, we can find

(i) holomorphic functions $f_1, \ldots, f_\ell$ on $\mathbb{C}^{n+N}$,

(ii) a connected Stein open neighborhood U ot t^0 in D ,

(iii) an open neighborhood B of $U \times \overline{\Delta^N(a')}$ in $U \times \Delta^N(b')$, and

(iv) $0 < \alpha < \beta$ in $\mathbb{R}$,

such that

(1) the holomorphic map $F: \mathbb{C}^{n+N} \longrightarrow \mathbb{C}^{n+\ell}$ defined by the coordinates of $\mathbb{C}^n$ and $f_1, \ldots, f_\ell$ makes $\tilde{V} \cap B \cap F^{-1}(U \times \Delta^\ell(\beta))$ an analytic cover over $U \times \Delta^\ell(\beta)$ and

(2) $\tilde{V} \cap (U \times \overline{\Delta^N(a')}) \subset F^{-1}(U \times \Delta^\ell(\alpha))$.

Let

$$\pi: \tilde{V} \cap B \cap F^{-1}(U \times G^\ell(\alpha,\beta)) \longrightarrow U \times G^\ell(\alpha,\beta) ,$$

$$\tilde{\pi}: \tilde{V} \cap B \cap F^{-1}(U \times \Delta^\ell(\beta)) \longrightarrow U \times \Delta^\ell(\beta)$$

be induced by F . Let

$$\mathcal{F}' = \mathcal{F} | \tilde{V} \cap B \cap F^{-1}(U \times G^\ell(\alpha,\beta))$$

and

$$\mathcal{G}' = \mathcal{G} \,|\, \tilde{V} \cap B \cap F^{-1}(U \times \Delta^{\ell}(\beta)) .$$

The zeroth direct image $\tilde{\pi}_*{}^0\mathcal{G}'$ of $\mathcal{G}'$ under π is a coherent analytic sheaf on $U \times \Delta^{\ell}(\beta)$. $\pi_*{}^0\mathcal{F}'$ is a coherent analytic subsheaf of $\tilde{\pi}_*{}^0\mathcal{G}' \,|\, U \times G^{\ell}(\alpha,\beta)$. For $t \in U$, $\tilde{\pi}_*{}^0(\mathcal{G}'(t)) = (\tilde{\pi}_*{}^0\mathcal{G}')(t)$ and $\pi_*{}^0(\mathcal{F}'(t)) = (\pi_*{}^0\mathcal{F}')(t)$.

After shrinking U and β if necessary, we can assume without loss of generality that there exists a sheaf-epimorphism

$$\varphi\colon {}_{n+\ell}\mathcal{O}^p \longrightarrow \tilde{\pi}_*{}^0\mathcal{G}'$$

on $U \times \Delta^{\ell}(\beta)$. Let $\mathcal{A} = \varphi^{-1}(\pi_*{}^0\mathcal{F}')$ on $U \times G^{\ell}(\alpha,\beta)$. Since $\mathcal{F}_{[k]\mathcal{G}} = \mathcal{F}$,

$$\mathcal{A}_{[n+\ell-1]_{n+\ell}\mathcal{O}^p} = \mathcal{A} .$$

For $t \in U$,

$$\operatorname{Im}\left(\mathcal{A}(t) \longrightarrow {}_{n+\ell}\mathcal{O}^p(t)\right) = \varphi(t)^{-1}\Big(\operatorname{Im}\big((\pi_*{}^0\mathcal{F}')(t) \longrightarrow (\tilde{\pi}_*{}^0\mathcal{G}')(t)\big)\Big),$$

where

$$\varphi(t)\colon {}_{n+\ell}\mathcal{O}^p(t) \longrightarrow (\tilde{\pi}_*{}^0\mathcal{G}')(t)$$

is induced by φ . It follows that for $t \in U \cap A$, $\operatorname{Im}\left(\mathcal{A}(t) \longrightarrow {}_{n+\ell}\mathcal{O}^p(t)\right)$ can be extended to a coherent analytic subsheaf of ${}_{n+\ell}\mathcal{O}^p(t) \,|\, U \times \Delta^{\ell}(\beta)$.

There is a decomposition ${}_{n+\ell}\mathcal{O}^p = {}_{n+\ell}\mathcal{O}^r \oplus {}_{n+\ell}\mathcal{O}^{p-r}$ such that the restriction $\psi\colon {}_{n+\ell}\mathcal{O}^r \longrightarrow {}_{n+\ell}\mathcal{O}^p/\mathcal{A}$ of the quotient map ${}_{n+\ell}\mathcal{O}^p \longrightarrow {}_{n+\ell}\mathcal{O}^p/\mathcal{A}$ to the first summand is an isomorphism at some point of

$$U \times G^{\ell}(\alpha,\beta) - S_{n+\ell-1}({}_{n+\ell}\mathcal{O}^p/\mathcal{A}).$$

By (2.A.7) there exists a thin set C in U such that $\dim(\operatorname{Coker} \psi)(t) \leq \ell-1$ for $t \in U-C$. For $t \in U-C$ the set of associated meromorphic vector-functions of $\operatorname{Im}(\mathcal{A}(t) \longrightarrow {}_{n+\ell}\mathcal{O}^p(t))$ (defined by

$${}_{n+\ell}\mathcal{O}^r(t) \longrightarrow {}_{n+\ell}\mathcal{O}^p(t)\Big/\Big(\operatorname{Im} \mathcal{A}(t) \longrightarrow {}_{n+\ell}\mathcal{O}^p(t)\Big)$$

induced by ψ) equals the restriction to $\{t\} \times G^{\ell}(\alpha,\beta)$ of the set of associated meromorphic vector-functions of $\mathcal{A}$ (defined by ψ). By (1.1) and (4.3) (and the remark following (4.3)), $\mathcal{A}$ can be extended to a coherent analytic subsheaf of ${}_{n+\ell}\mathcal{O}^p | U \times \Delta^{\ell}(\beta)$. It follows that $\pi_*{}^0\mathcal{F}'$ can be extended to a coherent analytic subsheaf of $\tilde{\pi}_*{}^0\mathcal{G}'$ on $U \times \Delta^{\ell}(\beta)$. $\pi_*{}^0\mathcal{F}'$ is generated on $U \times G^{\ell}(\alpha,\beta)$ by a subset of $\Gamma(U \times \Delta^{\ell}(\beta), \tilde{\pi}_*{}^0\mathcal{G}')$. Hence $\mathcal{F}'$ is generated by a subset of $\Gamma(\tilde{V} \cap B \cap F^{-1}(U \times \Delta^{\ell}(\beta)), \mathcal{G})$. $\mathcal{F}'$ can be extended to a coherent analytic subsheaf $\mathcal{F}^*$ of $\mathcal{G}'$ on $\tilde{V} \cap B \cap F^{-1}(U \times \Delta^{\ell}(\beta))$.

Let $\mathcal{F}^{\#}$ be the subsheaf of $\mathcal{G}$ on $\tilde{V} \cap (U \times \overset{N}{\Delta}(b))$ which agrees with $\mathcal{F}^*$ on $\tilde{V} \cap B \cap F^{-1}(U \times \Delta^{\ell}(\beta))$ and agrees with $\mathcal{F}$ on

$$\tilde{V} \cap (U \times \overset{N}{\Delta}(b)) - \pi^{-1}(U \times \overline{\Delta^{\ell}(\alpha)}).$$

$\mathcal{F}^{\#}$ is coherent. Let $\tilde{\mathcal{F}} = (\mathcal{F}^{\#})_{[k]\mathcal{G}}$. Since $\tilde{\mathcal{F}}$ agrees with $\mathcal{F}$ on $U \times \overset{N}{G}(b',b)$, by (4.4) $\tilde{\mathcal{F}}$ agrees with $\mathcal{F}$ on $U \times \overset{N}{G}(a,b)$.

(b) Assume that $\dim \tilde{V}(t) \leq k+1-n$ for $t \in D$. Let D'

be the largest open subset D such that $\mathcal{F}|D' \times G^N(a,b)$ can be extended to a coherent analytic subsheaf of $\mathcal{G}|D' \times \triangle^N(b)$. It follows from (a) that D' is a nonempty closed subset of D . Hence $D' = D$.

(c) Let $\sigma: \tilde{V} \longrightarrow D$ be induced by the natural projection $D \times \triangle^N(b) \longrightarrow D$. Let S be the topological closure of the set of points of $\tilde{V}$ where the rank of σ is $< n$. Take $a < b' < b$ in $\mathbb{R}^N$. By (2.A.7), $\sigma(S \cap (D \times \overline{\triangle^N(b')}))$ is a closed thin set in D . Let $D' = D - \sigma(S \cap (D \times \overline{\triangle^N(b')}))$. By (b), $\mathcal{F}|D' \times G^N(a,b)$ can be extended to a coherent analytic subsheaf of $\mathcal{G}|D' \times \triangle^N(b)$.

Let L be an arbitrary relatively compact open subset of D . By (2.2) there exists an N-dimensional plane T in $\mathbb{C}^n \times \mathbb{C}^N$ such that for some nonempty open subset Q of L and for some connected open neighborhood R of L in D we have

(i) $(Q+T) \cap (\mathbb{C}^n \times \triangle^N(b)) \subset D' \times \triangle^N(b)$

(ii) $L \times \triangle^N(b) \subset (R+T) \cap (\mathbb{C}^n \times \triangle^N(b)) \subset D \times \triangle^N(b)$

(iii) $\dim(x+T) \cap \tilde{V} \leq k+l-n$ for $x \in V$,

where

$$Q+T = \{x+y \mid x \in Q,\ y \in T\}$$

and

$$x+T = \{x+y \mid y \in T\} .$$

By (b), $\mathcal{F}|(R+T) \cap (\mathbb{C}^n \times G^N(a,b))$ can be extended to a coherent analytic subsheaf $\mathcal{F}^*$ of $\mathcal{G}|(R+T) \cap (\mathbb{C}^n \times \triangle^N(b))$. It follows that $\mathcal{F}^*|L \times \triangle^N(b)$ is a coherent analytic subsheaf of $\mathcal{G}|L\times\Delta^N(b)$ extending $\mathcal{F}|L\times G^N(a,b)$. Since L is an arbitrary relatively compact open subset of D , $\mathcal{F}$ can be extended to a coherent analytic subsheaf of $\mathcal{G}$ on $D\times\Delta^N(b)$.

Q.E.D.

(4.6) Corollary. Suppose $0 \le a < b$ in $\mathbb{R}^N$, D is an open subset of $\mathbb{C}^n$, $\mathcal{G}$ is a coherent analytic sheaf on $D\times\Delta^N(b)$, and $\mathcal{F}$ is a coherent analytic subsheaf of $\mathcal{G}$ on $D\times G^N(a,b)$ such that $\mathcal{F}_{[n+1]\mathcal{G}} = \mathcal{F}$. Then $\mathcal{F}$ can be extended uniquely to a coherent analytic subsheaf $\tilde{\mathcal{F}}$ of $\mathcal{G}$ on $D\times\Delta^N(b)$ satisfying $\tilde{\mathcal{F}}_{[n]\mathcal{G}} = \tilde{\mathcal{F}}$.

Proof. Identify D with the subset $D\times\{0\}$ of $D\times\mathbb{C}$ and extend trivially $\mathcal{G}$ (respectively $\mathcal{F}$) to a coherent analytic sheaf $\tilde{\mathcal{G}}$ on $D\times\mathbb{C}\times\Delta^N(b)$ (respectively $\tilde{\mathcal{F}}$ on $D\times\mathbb{C}\times G^N(a,b)$). The corollary follows from applying (4.5) to the subsheaf $\tilde{\mathcal{F}}$ of $\tilde{\mathcal{G}}|D\times\mathbb{C}\times G^N(a,b)$. Q.E.D.

CHAPTER 5

EXTENSION OF LOCALLY FREE SHEAVES ON RING DOMAINS

In this chapter we prove the extension theorem for locally free sheaves on ring domains.

Suppose $\mathcal{F}$ is a locally free sheaf on a complex manifold $(M,\mathcal{O})$. Choose two open coverings $\{U_\alpha\}$, $\{\tilde{U}_\alpha\}$ of M indexed by the same set such that, for every α, $U_\alpha \subset\subset \tilde{U}_\alpha$ and there exists a sheaf-isomorphism $\varphi_\alpha : \mathcal{O}^r | \tilde{U}_\alpha \longrightarrow \mathcal{F} | \tilde{U}_\alpha$. If $\mathcal{V} = \{V_i\}$ is an open covering of an open subset of M such that every V_i is relatively compact in M, then we denote by $C^p_{L^2}(\mathcal{V},\mathcal{F})$ the set of all $\{\xi_{i_0\ldots i_p}\} \in C^p(\mathcal{V},\mathcal{F})$ such that, for every α,

$$\varphi_\alpha^{-1}\left(\xi_{i_0\ldots i_p}\Big| U_\alpha \cap V_{i_0} \cap \ldots \cap V_{i_p}\right)$$

is an r-tuple of L^2 functions on $U_\alpha \cap V_{i_0} \cap \ldots \cap V_{i_p}$ with respect to some C^∞ volume element ω_α on $\tilde{U}_\alpha$. Clearly $C^p_{L^2}(\mathcal{V},\mathcal{F})$ is independent of the choices of $\{U_\alpha\}$, $\{\tilde{U}_\alpha\}$, $\{\varphi_\alpha\}$, and $\{\omega_\alpha\}$. The kernel and the image of the map

$$C^p_{L^2}(\mathcal{V},\mathcal{F}) \longrightarrow C^{p+1}_{L^2}(\mathcal{V},\mathcal{F})$$

induced by the coboundary map are denoted respectively by $Z^p_{L^2}(\mathcal{V},\mathcal{F})$ and $B^{p+1}_{L^2}(\mathcal{V},\mathcal{F})$. Denote $Z^p_{L^2}(\mathcal{V},\mathcal{F}) \big/ B^p_{L^2}(\mathcal{V},\mathcal{F})$ by $H^p_{L^2}(\mathcal{V},\mathcal{F})$.

(5.1) Proposition. Suppose $0 \leq \tilde{a} < a^* \leq a < b \leq b^* < \tilde{b}$ in $\mathbb{R}^N$. Let $\mathcal{U} = \{U_\alpha\}_{\alpha=1}^N$ and $\mathcal{V} = \{V_j\}_{j=1}^N$, where
$U_\alpha = \Delta^N(b^*) \cap \{|z_\alpha| > a_\alpha^*\}$ and $V_j = \Delta^N(b) \cap \{|z_j| > a_j\}$.
Suppose $\mathcal{F}$ is a locally free sheaf on $G^N(\tilde{a},\tilde{b})$ and $N \geq 3$.
Then the restriction map $H^1_{L^2}(\mathcal{U},\mathcal{F}) \longrightarrow H^1_{L^2}(\mathcal{V},\mathcal{F})$ is surjective.
In particular, $\dim_{\mathbb{C}} H^1_{L^2}(\mathcal{V},\mathcal{F}) < \infty$.

Proof. To prove the surjectivity of

$$H^1_{L^2}(\mathcal{U},\mathcal{F}) \longrightarrow H^1_{L^2}(\mathcal{V},\mathcal{F}) ,$$

it suffices to prove the special case where $a_j^* = a_j$ and $b_j = b_j^*$ for $2 \leq j \leq N$, because the general case follows from repeating the special case N times.

Let $\mathcal{W} = \{W_j\}_{j=1}^N$, where $W_j = V_j$ for $2 \leq j \leq N$ and

$$W_1 = \Delta^N(b^*) \cap \{|z_1| > a_1\} .$$

Since $W_i \cap W_j = V_i \cap V_j$ for $i \neq j$, we have $Z^1_{L^2}(\mathcal{V},\mathcal{F}) = Z^1_{L^2}(\mathcal{W},\mathcal{F})$. It suffices to prove the surjectivity of $H^1_{L^2}(\mathcal{U},\mathcal{F}) \longrightarrow H^1_{L^2}(\mathcal{W},\mathcal{F})$.

There exists a sheaf-isomorphism $\varphi_\alpha : {}_N\mathcal{O}^r \longrightarrow \mathcal{F}$ on

$$\Delta^N(\tilde{b}) \cap \{|z_\alpha| > \tilde{a}_\alpha\}$$

for $1 \leq \alpha \leq N$ (see (5.A.8) of the Appendix of Chapter 5).

Fix $\xi = \{\xi_{\alpha\beta}\} \in Z^1_{L^2}(\mathcal{W},\mathcal{F})$. For $2 \leq \alpha \leq N$ let

$$\varphi_\alpha^{-1}(\xi_{\alpha 1}) = \sum_{k=-\infty}^{\infty} c_k^{(\alpha)} z_1^{\,k}$$

be the Laurent series expansion of $\varphi_\alpha^{-1}(\xi_{\alpha 1})$ in z_1 . Define

$$f_\alpha' = \sum_{k=0}^{\infty} c_k^{(\alpha)} z_1^{\,k} \in \Gamma(W_\alpha, {}_N\mathcal{O}^r)$$

and

$$g_\alpha' = \sum_{k=-\infty}^{-1} c_k^{(\alpha)} z_1^{\,k} \in \Gamma(W_1 \cap U_\alpha, {}_N\mathcal{O}^r) .$$

By (3.1), f_α' is an r-tuple of L^2 functions on W_α and g_α' is an r-tuple of L^2 functions on $W_1 \cap U_\alpha$. Let $f_\alpha = \varphi_\alpha(f_\alpha') \in \Gamma(W_\alpha, \mathcal{F})$ and $g_\alpha = \varphi_\alpha(g_\alpha') \in \Gamma(W_1 \cap U_\alpha, \mathcal{F})$.

Consider the covering

$$\mathcal{E} := \{W_j \cap U_1\}_{j=1}^{N}$$

of $U_1 \cap G^N(a,b)$. Let

$$\gamma = \{\gamma_{ij}\} \in Z^1_{L^2}(\mathcal{E}, {}_N\mathcal{O}^r)$$

be defined by

$$\gamma_{ij} = \varphi_1^{-1}(\xi_{ij} | W_i \cap W_j \cap U_1) .$$

By (3.2) there exists

$$\lambda' = \{\lambda_i'\} \in C^0_{L^2}(\mathcal{E}, {}_N\mathcal{O}^r)$$

such that $\delta\lambda' = \gamma$. Let $\lambda_1' = \sum_{k=-\infty}^{\infty} d_k z_1^{\,k}$ be the Laurent series expansion of λ_1' in z_1 . Define

$$\lambda_1^* = \sum_{k=-\infty}^{-1} d_k z_1^k \in \Gamma(W_1, {}_N\mathcal{O}^r)$$

and

$$\lambda_j^* = \lambda_j' - \sum_{k=0}^{\infty} d_k z_1^k \in \Gamma(W_j \cap U_1, {}_N\mathcal{O}^r)$$

for $2 \leq j \leq N$. By (3.1) λ_j^* is an r-tuple of L^2 functions on $W_i \cap U_1$ for $1 \leq j \leq N$. Let

$$\lambda_j = \varphi_1(\lambda_j^*) \in \Gamma(W_j \cap U_1, \mathcal{F})$$

for $1 \leq j \leq N$. Then

$$(*) \qquad \xi_{ij} = \lambda_j - \lambda_i \quad \text{on} \quad W_i \cap W_j \cap U_1 .$$

For $1 \leq i, \alpha \leq N$ define

$$\sigma_{i,\alpha} \in \Gamma(W_i \cap U_\alpha, \mathcal{F})$$

by

$$\sigma_{i,\alpha} = \begin{cases} \xi_{\alpha i} - f_\alpha & \text{for } 2 \leq i, \alpha \leq N \\ g_\alpha & \text{for } i = 1 \text{ and } 2 \leq \alpha \leq N \\ \lambda_i & \text{for } 1 \leq i \leq N \text{ and } \alpha = 1 . \end{cases}$$

It follows from $\delta\xi = 0$, $\xi_{\alpha 1} = f_\alpha + g_\alpha$, and (*) that

$$(\dagger) \qquad \xi_{ij} = \sigma_{j,\alpha} - \sigma_{i,\alpha} \quad \text{on} \quad W_i \cap W_j \cap U_\alpha .$$

Define

$$\tau_{\alpha\beta} \in \Gamma(U_\alpha \cap U_\beta, \mathcal{F})$$

by

$$\tau_{\alpha\beta} | U_\alpha \cap U_\beta \cap W_i = (\sigma_{i,\alpha} - \sigma_{i,\beta}) | U_\alpha \cap U_\beta \cap W_i .$$

Because of (†), $\tau_{\alpha\beta}$ is well-defined. Let $\tau = \{\tau_{\alpha\beta}\}$. Then $\tau \in Z^1_{L^2}(\mathfrak{U}, \mathfrak{F})$. Define

$$\theta = \{\theta_i\} \in C^0_{L^2}(\mathfrak{W}, \mathfrak{F})$$

by $\theta_i = \sigma_{i,i}$ $(1 \leq i \leq N)$. Let

$$\rho: Z^1_{L^2}(\mathfrak{U}, \mathfrak{F}) \longrightarrow Z^1_{L^2}(\mathfrak{W}, \mathfrak{F})$$

be the restriction map. Then $\rho(\tau) = \xi - \delta\theta$, because, on $W_i \cap W_j$,

$$\begin{aligned} 2(\tau_{ij} - \xi_{ij}) &= (\sigma_{i,i} - \sigma_{i,j}) + (\sigma_{j,i} - \sigma_{j,j}) - (\sigma_{j,i} - \sigma_{i,i}) - (\sigma_{j,j} - \sigma_{i,j}) \\ &= 2(\sigma_{i,i} - \sigma_{j,j}) = 2(\theta_i - \theta_j) . \end{aligned}$$

The surjectivity of

$$H^1_{L^2}(\mathfrak{U}, \mathfrak{F}) \longrightarrow H^1_{L^2}(\mathfrak{W}, \mathfrak{F})$$

is proved.

To prove the finite-dimensionality of $H^1_{L^2}(\mathfrak{V}, \mathfrak{F})$, we assume that $a^* < a$ and $b < b^*$. Let

$$r: Z^1_{L^2}(\mathfrak{U}, \mathfrak{F}) \longrightarrow Z^1_{L^2}(\mathfrak{V}, \mathfrak{F})$$

be the restriction map. Define

$$\Psi, \chi: Z^1_{L^2}(\mathfrak{U}, \mathfrak{F}) \oplus C^0_{L^2}(\mathfrak{V}, \mathfrak{F}) \longrightarrow Z^1_{L^2}(\mathfrak{V}, \mathfrak{F})$$

by

$$\Psi(a \oplus b) = r(a) + \delta b$$

and $\chi(a \oplus b) = r(a)$. Since ψ is surjective and χ is compact, $\dim \operatorname{Coker}(\psi - \chi) < \infty$. Since $H^1_{L^2}(\mathcal{V}, \mathcal{F})$ is isomorphic to $\operatorname{Coker}(\psi - \chi)$, $\dim H^1_{L^2}(\mathcal{V}, \mathcal{F}) < \infty$. Q.E.D.

Suppose D is an open subset of $\mathbb{C}^n$ and E, F are Banach spaces. $L(E,F)$ denotes the Banach space of all continuous linear maps from E to F . $\operatorname{Hol}(D,E)$ denotes the space of all holomorphic maps from D to E furnished with the topology of uniform convergence on compact subsets. A map $\varphi: D \times E \longrightarrow D \times F$ is called a *holomorphic bundle-homomorphism* if there exists $\varphi' \in \operatorname{Hol}(D, L(E,F))$ such that $\varphi(x,e) = (x, \varphi'(x)\, e)$. For notational simplicity we use φ also to denote φ' .

The use of linear algebra in the proofs of the following proposition and (6.3) is inspired by [12].

(5.2) **Proposition.** *Suppose* E, F, G *are Banach spaces,* D *is an open neighborhood of* 0 *in* $\mathbb{C}^n$ *, and* $D \times E \xrightarrow{\varphi} D \times F \xrightarrow{\psi} D \times G$ *is a complex of holomorphic bundle-homomorphisms. Suppose* $\dim_{\mathbb{C}} \operatorname{Ker} \psi(0) / \operatorname{Im} \varphi(0) < \infty$ *and* $\operatorname{Ker} \varphi(0)$ *and* $\operatorname{Ker} \psi(0)$ *are complemented respectively in* E *and* F *. Then there exists an open neighborhood* U *of* 0 *in* D *such that, for every connected Stein open neighborhood* W *of* 0 *in* U *, the image of the map* $\varphi_W: \operatorname{Hol}(W,E) \longrightarrow \operatorname{Hol}(W,F)$ *induced by* φ *is closed.*

Proof. (a) Consider first the special case where $F = \mathbb{C}^p$ and $G = 0$. Let U be a relatively compact open neighborhood of

0 in D . We claim that U satisfies the requirement.

For $e \in E$ let

$$\sigma(e) \in \Gamma(D, {}_n\mathcal{O}^p) = \mathrm{Hol}(D, \mathbb{C}^p)$$

be the image under φ_D of the element of Hol(D,E) which maps D to $\{e\}$. Let $\mathcal{A}$ be the subsheaf of ${}_n\mathcal{O}^p|D$ generated by $\{\sigma(e)\}_{e \in E}$. Since $\mathcal{A}$ is coherent, there exists a finite subset $\{e_1, \ldots, e_q\}$ of E such that $\sigma(e_1), \ldots, \sigma(e_q)$ generate $\mathcal{A}|U$.

Suppose W is a Stein open subset of U and suppose $f_\nu, f \in \Gamma(W, {}_n\mathcal{O}^p)$ $(1 \leq \nu < \infty)$ such that $f_\nu \longrightarrow f$ and $f_\nu = \varphi_W(g_\nu)$ for some $g_\nu \in \mathrm{Hol}(W,E)$. We are going to show that $f_t \in \mathcal{A}_t$ for $t \in W$.

Fix $t^0 \in W$. Take an open polydisc P in W, with radius $r > 0$, centered at t^0. For every $1 \leq \nu < \infty$ we have a power series expansion

$$g_\nu(t) = \sum_\alpha e_{\nu,\alpha}(t-t^0)^\alpha \qquad \text{for } t \in P,$$

where $\alpha = (\alpha_1, \ldots, \alpha_n)$ and

$$(t-t^0)^\alpha = (t_1-t_1^0)^{\alpha_1} \ldots (t_n-t_n^0)^{\alpha_n}.$$

Since

$$f_\nu(t) = \sum_\alpha \sigma(e_{\nu,\alpha})(t-t^0)^\alpha \qquad \text{for } t \in P,$$

it follows from $\sigma(e_{\alpha,\nu}) \in \Gamma(D,\mathcal{A})$ that $(f_\nu)_{t^0} \in \mathcal{A}_{t^0}$. Hence $f_{t^0} \in \mathcal{A}_{t^0}$.

Since W is Stein, $f = \sum_{i=1}^{q} a_i \sigma(e_i)$ on W for some $a_i \in \Gamma(W, {}_n\mathcal{O})$. It follows that $f = \varphi_W\left(\sum_{i=1}^{q} a_i e_i | W\right)$. Hence $\operatorname{Im} \varphi_W$ is closed.

(b) For the general case, we have the following decomposition of Banach spaces into direct sums of closed subspaces:

$$E = E_1 \oplus E_2$$

$$F = F_1 \oplus F_2 \oplus F_3 ,$$

where $E_2 = \operatorname{Ker} \varphi(0)$, $F_1 = \operatorname{Im} \varphi(0)$, and $F_1 \oplus F_2 = \operatorname{Ker} \psi(0)$. φ and ψ can be written as

$$\varphi = \begin{pmatrix} \varphi_{11} & \varphi_{12} \\ \varphi_{21} & \varphi_{22} \\ \varphi_{31} & \varphi_{32} \end{pmatrix}$$

$$\psi = (\psi_1 \quad \psi_2 \quad \psi_3) ,$$

where φ_{ij} is a holomorphic bundle-homomorphism from $D \times E_j$ to $D \times F_i$ and ψ_i is a holomorphic bundle-homomorphism from $D \times F_i$ to $D \times G$.

Since $\varphi_{11}(0)$ is an isomorphism from E_1 to F_1, there exists an open neighborhood $\tilde{U}$ of 0 in D such that, for $t \in \tilde{U}$, $\varphi_{11}(t)$ is an isomorphism from E_1 to F_1. Let $\theta : \tilde{U} \times E_2 \longrightarrow \tilde{U} \times F_2$ be the holomorphic bundle-homomorphism defined by $\varphi_{22} - \varphi_{21}\varphi_{11}^{-1}\varphi_{12}$. Since $\dim_{\mathbb{C}} F_2 < \infty$, by (a) there exists an open neighborhood U of 0 in $\tilde{U}$ such that, for every connected Stein open neighborhood W of 0 in U, the image of θ_W: $\mathrm{Hol}(W,E_2) \longrightarrow \mathrm{Hol}(W,F_2)$ induced by θ is closed. We claim that U satisfies the requirement for the general case.

Suppose W is a connected Stein open neighborhood of 0 in U. Suppose

$$f_1^{(\nu)} \oplus f_2^{(\nu)} \oplus f_3^{(\nu)},\ f_1 \oplus f_2 \oplus f_3 \ \in\ \mathrm{Hol}(W,F)$$

$(1 \leq \nu < \infty)$ such that $f_i^{(\nu)} \longrightarrow f_i$ and

$$f_1^{(\nu)} \oplus f_2^{(\nu)} \oplus f_3^{(\nu)} \ \in\ \mathrm{Im}\ \varphi_W\ ,$$

where $f_i^{(\nu)}$, $f_i \in \mathrm{Hol}(W,F_i)$. We have to show that $f_1 \oplus f_2 \oplus f_3 \ \in\ \mathrm{Im}\ \varphi_W$.

There exist $e_j^{(\nu)} \in \mathrm{Hol}(W,E_j)$ $(1 \leq \nu < \infty,\ j = 1,2)$ such that

$$(*) \qquad \begin{cases} f_1^{(\nu)} = \varphi_{11}e_1^{(\nu)} + \varphi_{12}e_2^{(\nu)} \\ f_2^{(\nu)} = \varphi_{21}e_1^{(\nu)} + \varphi_{22}e_2^{(\nu)} \\ f_3^{(\nu)} = \varphi_{31}e_1^{(\nu)} + \varphi_{32}e_3^{(\nu)} \ . \end{cases}$$

By eliminating $e_1^{(\nu)}$ from the first two equations of (*), we obtain

$$f_2^{(\nu)} - \varphi_{21}\varphi_{11}^{-1}f_1^{(\nu)} = (\varphi_{22}-\varphi_{21}\varphi_{11}^{-1}\varphi_{12})e_2^{(\nu)} = \theta\, e_2^{(\nu)}\ .$$

Since

$$f_2^{(\nu)} - \varphi_{21}\varphi_{11}^{-1}f_1^{(\nu)} \ \in\ \mathrm{Im}\ \theta_W$$

and

$$f_2^{(\nu)} - \varphi_{21}\varphi_{11}^{-1}f_1^{(\nu)} \longrightarrow f_2 - \varphi_{21}\varphi_{11}^{-1}f_1\ ,$$

we have

$$f_2 - \varphi_{21}\varphi_{11}^{-1}f_1 \in \operatorname{Im}\theta_W .$$

There exists $e_2 \in \mathrm{Hol}(W,E_2)$ such that

$$f_2 - \varphi_{21}\varphi_{11}^{-1}f_1 = (\varphi_{22}-\varphi_{21}\varphi_{11}^{-1}\varphi_{12})e_2 .$$

Define

$$e_1 = \varphi_{11}^{-1}f_1 - \varphi_{11}^{-1}\varphi_{12}e_2 \in \mathrm{Hol}(W,E_1) .$$

Then

$$\begin{cases} f_1 = \varphi_{11}e_1 + \varphi_{12}e_2 \\ f_2 = \varphi_{21}e_1 + \varphi_{22}e_2 . \end{cases}$$

Let $g = f_3-(\varphi_{31}e_1 + \varphi_{32}e_2)$. We have to show that $g = 0$.

Since

$$\psi(0 \oplus 0 \oplus g) = \psi(f_1 \oplus f_2 \oplus f_3 - \varphi(e_1 \oplus e_2)) = 0 ,$$

we have $\psi_3 g = 0$. Suppose $g \neq 0$. Let k be the smallest integer such that some partial derivative

$$\left(\frac{\partial^k g}{\partial t_1^{\alpha_1} \cdots \partial t_n^{\alpha_n}}\right)(0)$$

of g of order k at 0 is nonzero, where $\alpha_1 + \dots + \alpha_n = k$. Then

$$\left(\frac{\partial^k \psi_3 g}{\partial t_1^{\alpha_1} \cdots \partial t_n^{\alpha_n}}\right)(0) = \psi_3(0)\left(\frac{\partial^k g}{\partial t_1^{\alpha_1} \cdots \partial t_n^{\alpha_n}}\right)(0) .$$

Since $\psi_3(0)$ is injective, this contradicts $\psi_3 g = 0$. Hence

$g = 0$ and

$$f_1 \oplus f_2 \oplus f_3 = \varphi(e_1 \oplus e_2) \in \operatorname{Im} \varphi_W . \qquad \text{Q.E.D.}$$

Suppose $0 \leq \tilde{a} < a < b < \tilde{b}$ in $\mathbb{R}^N$ and D is an open polydisc neighborhood of 0 in $\mathbb{C}^n$. Let

$$\tilde{U}_j = \Delta^N(\tilde{b}) \cap \{|z_j| > \tilde{a}_j\}$$

$(1 \leq j \leq N)$. Suppose $\mathcal{F}$ is a locally free sheaf on $D \times G^N(\tilde{a},\tilde{b})$. Then there exists a sheaf-isomorphism $\alpha_j \colon {}_{n+N}\mathcal{O}^r \longrightarrow \mathcal{F}$ on $D \times \tilde{U}_j$ for $1 \leq j \leq N$ (see the Appendix of Chapter 5). Let $\mathcal{U} = \{U_j\}_{j=1}^N$, where

$$U_j = \Delta^N(b) \cap \{|z_j| > a_j\} .$$

If W is an open subset of D, then we use the following notations. $W \times \mathcal{U}$ denotes $\{W \times U_j\}_{j=1}^N$. $C^p_{L^2_v}(W \times \mathcal{U}, \mathcal{F})$ denotes the set of all

$$\{\xi_{i_0 \ldots i_p}\} \in C^p(W \times \mathcal{U}, \mathcal{F})$$

such that

$$t \longmapsto \alpha_{i_0}^{-1}(\xi_{i_0 \ldots i_p})(t, \cdot)$$

is a holomorphic map from W to the Hilbert space of all r-tuples of holomorphic L^2 functions on $U_{i_1} \cap \ldots \cap U_{i_p}$. $H^p_{L^2_v}(W \times \mathcal{U}, \mathcal{F})$ denotes the pth cohomology group of the cochain complex

$$\ldots \longrightarrow C^{p-1}_{L^2_v}(W \times \mathcal{U}, \mathcal{F}) \longrightarrow C^p_{L^2_v}(W \times \mathcal{U}, \mathcal{F}) \longrightarrow C^{p+1}_{L^2_v}(W \times \mathcal{U}, \mathcal{F}) \longrightarrow \ldots .$$

$H^p_{L^2_v}(W \times \mathfrak{U}, \mathcal{F})$ is given the natural quotient topology.

(5.3) Proposition. *If $N \geq 3$, then there exists an open neighborhood U of 0 in D such that $H^1_{L^2_v}(W \times \mathfrak{U}, \mathcal{F})$ is Hausdorff for every connected Stein open neighborhood W of 0 in U.*

Proof. For $t \in D$ and $1 \leq j \leq N$ let $\alpha_j(t): {}_{n+N}\mathcal{O}^r(t) \longrightarrow \mathcal{F}(t)$ be induced by α_j and we identify ${}_{n+N}\mathcal{O}^r(t)$ with ${}_N\mathcal{O}^r$.

Let $H_{i_0 \cdots i_p}$ be the Hilbert space of all r-tuples of holomorphic L^2 functions on $U_{i_0} \cap \cdots \cap U_{i_p}$. Set

$$E = \bigoplus_{1 \leq i \leq N} H_i ,$$

$$F = \bigoplus_{1 \leq i < j \leq N} H_{ij} ,$$

and

$$G = \bigoplus_{1 \leq i < j < k \leq N} H_{ijk} .$$

Define a complex of holomorphic bundle-homomorphisms

$$D \times E \xrightarrow{\varphi} D \times F \xrightarrow{\psi} D \times G$$

as follows. For $t \in D$ and $e = \{e_i\} \in E$ define $\varphi(t,e) = (t,f)$, where $f = \{f_{ij}\}$ and

$$f_{ij} = \alpha_i(t)^{-1}\left(\alpha_j(t)e_j - \alpha_i(t)e_i\right) .$$

For $t \in D$ and $f = \{f_{ij}\} \in F$ define $\psi(t,f) = (t,g)$, where

$g = \{g_{ijk}\}$ and

$$g_{ijk} = \alpha_i(t)^{-1}\left(\alpha_j(t)f_{jk} - \alpha_i(t)f_{ik} + \alpha_i(t)f_{ij}\right).$$

For $t \in D$, $\operatorname{Ker} \psi(t)/\operatorname{Im} \varphi(t)$ is isomorphic to $H^1_{L^2}(\mathfrak{U},\mathcal{F}(t))$ which is finite-dimensional by (5.1). For any open subset W of D, $\operatorname{Ker} \psi_W/\operatorname{Im} \varphi_W$ furnished with the natural quotient topology is topologically isomorphic to $H^1_{L^2_v}(W \times \mathfrak{U},\mathcal{F})$, where

$$\varphi_W\colon \operatorname{Hol}(W,E) \longrightarrow \operatorname{Hol}(W,F)$$

and

$$\psi_W\colon \operatorname{Hol}(W,F) \longrightarrow \operatorname{Hol}(W,G)$$

are induced respectively by φ and ψ. The proposition now follows from (5.2). Q.E.D.

Suppose X is a complex space (with countable topology) and $\mathcal{F}$ is a coherent analytic sheaf on X. We furnish $H^p(X,\mathcal{F})$ with a topology in the following way. Take a countable Stein covering $\mathfrak{U}$ of X. The topology of $H^p(X,\mathcal{F})$ is defined as the quotient topology of $Z^p(\mathfrak{U},\mathcal{F})/B^p(\mathfrak{U},\mathcal{F})$, where $Z^p(\mathfrak{U},\mathcal{F})$ is given the natural Fréchet space topology. This topology of $H^p(X,\mathcal{F})$ is independent of the choice of $\mathfrak{U}$. We denote $H^p(X,\mathcal{F})/\{0\}^-$ by $\overline{H}^p(X,\mathcal{F})$, where $\{0\}^-$ is the topological closure of 0 in $H^p(X,\mathcal{F})$.

If X_1, X_2 are open subsets of X such that $X = X_1 \cup X_2$ and $(X_1-X_2)^- \cap (X_2-X_1)^- = \emptyset$, then the maps in the following Mayer-Vietoris sequence are all continuous:

$$\cdots \to H^{p-1}(X_1 \cap X_2, \mathcal{F}) \to H^p(X, \mathcal{F}) \to$$

$$H^p(X_1, \mathcal{F}) \oplus H^p(X_2, \mathcal{F}) \to H^p(X_1 \cap X_2, \mathcal{F}) \to \cdots$$

(see the Appendix of Chapter 5 for proofs of these assertions).

(5.4) Lemma. *Suppose* $\mathcal{F}$ *is a coherent analytic sheaf on a complex space* X *and* X_1, X_2 *are open subsets of* X *such that* $X_1 \cup X_2 = X$ *and* $(X_1 - X_2)^- \cap (X_2 - X_1)^- = \emptyset$. *Suppose* $p \geq 1$, $H^p(X_1 \cap X_2, \mathcal{F}) = 0$, *and* $\mathrm{Im}\left(H^{p-1}(X_2, \mathcal{F}) \to H^{p-1}(X_1 \cap X_2, \mathcal{F})\right)$ *is dense in* $H^{p-1}(X_1 \cap X_2, \mathcal{F})$. *Then the restriction map* $\theta : H^p(X, \mathcal{F}) \to H^p(X_1, \mathcal{F})$ *is surjective and it induces a topological isomorphism* $\bar{H}^p(X, \mathcal{F}) \to \bar{H}^p(X_1, \mathcal{F})$.

Proof. Let $X_{12} = X_1 \cap X_2$. Consider the following portion of the Mayer-Vietoris sequence

$$H^{p-1}(X_1, \mathcal{F}) \oplus H^{p-1}(X_2, \mathcal{F}) \xrightarrow{\alpha} H^{p-1}(X_{12}, \mathcal{F}) \xrightarrow{\beta}$$

$$H^p(X, \mathcal{F}) \xrightarrow{\gamma} H^p(X_1, \mathcal{F}) \oplus H^p(X_2, \mathcal{F}) \to H^p(X_{12}, \mathcal{F}).$$

Since $H^p(X_{12}, \mathcal{F}) = 0$, γ is surjective. In particular, θ is surjective. Let N (respectively N_1) be the topological closure of $\{0\}$ in $H^p(X, \mathcal{F})$ (respectively in $H^p(X_1, \mathcal{F})$). Since $\bar{H}^p(X, \mathcal{F})$ and $\bar{H}^p(X_1, \mathcal{F})$ are Fréchet spaces, to finish the proof, we need only show that $\theta^{-1}(N_1) \subset N$. Suppose G is an open subset of $H^p(X, \mathcal{F})$ intersecting $\theta^{-1}(N_1)$. It suffices to show that $0 \in G$. Choose a countable Stein open covering $\mathcal{U}$ of X such that

$$\mathfrak{U}_1 := \{U \in \mathfrak{U} \mid U \subset X_1\}$$

covers X_1 . Since θ is surjective, the restriction map

$$Z^p(\mathfrak{U},\mathcal{F}) \longrightarrow Z^p(\mathfrak{U}_1,\mathcal{F})$$

is surjective and hence is open. It follows that θ is open. Since $G \cap \theta^{-1}(N_1) \neq \emptyset$, $\theta(G)$ is an open subset of $H^p(X_1,\mathcal{F})$ intersecting N_1 . Hence $0 \in \theta(G)$ and $\beta^{-1}(G)$ is a non-empty open subset of $H^{p-1}(X_{12},\mathcal{F})$. Since $\operatorname{Im}\alpha$ is dense in $H^{p-1}(X_{12},\mathcal{F})$, $(\operatorname{Im}\alpha) \cap \beta^{-1}(G) \neq \emptyset$. It follows that $0 \in G$.

Q.E.D.

(5.5) Proposition. *Suppose* $a^* \leq a < b \leq b^*$ *in* $\mathbb{R}^N$, D *is a Stein open subset of* $\mathbb{C}^n$, *and* $\mathcal{F}$ *is a locally free sheaf on* $D \times G^N(a^*,b^*)$. *If* $N \geq 3$, *then the restriction map* $\theta: H^1(D \times G^N(a^*,b^*),\mathcal{F}) \longrightarrow H^1(D \times G^N(a,b),\mathcal{F})$ *is surjective and it induces a topological isomorphism* $\overline{H}^1(D \times G^N(a^*,b^*),\mathcal{F}) \longrightarrow \overline{H}^1(D \times G^N(a,b),\mathcal{F})$.

Proof. It suffices to prove the special case where $a_j^* = a_j$ and $b_j = b_j^*$ for $2 \leq j \leq N$, because the general case follows from applying the special case N times.

(a) Assume first that $b_1 = b_1^*$. We are going to show that θ is an isomorphism. Let $Y_1 = D \times G^N(a,b)$ and

$$Y_2 = D \times (\Delta^N(b) \cap \{|z_1| > a_1^*\}) .$$

Then

$$Y_1 \cup Y_2 = D \times G^N(a^*,b^*)$$

and

$$Y_1 \cap Y_2 = D \times (G^N(a,b) \cap \{|z_1| > a_1^*\}) .$$

By (3.2), $H^1(Y_1 \cap Y_2, {}_{n+N}\mathcal{O}) = 0$. By (1.2),

$$\Gamma(Y_2, {}_{n+N}\mathcal{O}) \longrightarrow \Gamma(Y_1 \cap Y_2, {}_{n+N}\mathcal{O})$$

is surjective. Since $\mathcal{F} \approx {}_{n+N}\mathcal{O}^r$ on Y_2 for some r , it follows from the following portion of the Mayer-Vietoris sequence

$$\Gamma(Y_1,\mathcal{F}) \oplus \Gamma(Y_2,\mathcal{F}) \longrightarrow \Gamma(Y_1 \cap Y_2,\mathcal{F}) \longrightarrow H^1(Y_1 \cup Y_2,\mathcal{F})$$

$$\longrightarrow H^1(Y_1,\mathcal{F}) \oplus H^1(Y_2,\mathcal{F}) \longrightarrow H^1(Y_1 \cap Y_2,\mathcal{F})$$

that θ is an isomorphism.

(b) To finish the proof, we can assume that $a_1 = a_1^*$. Let $X_1 = D \times G^N(a,b)$ and

$$X_2 = D \times (\Delta^N(b^*) \cap \{|z_1| > a_1\}) .$$

Then

$$X_1 \cup X_2 = D \times G^N(a^*,b^*)$$

and

$$X_1 \cap X_2 = D \times (\Delta^N(b) \cap \{|z_1| > a_1\}) .$$

We have

$$(X_1 - X_2)^- \cap (X_2 - X_1)^- = \emptyset .$$

Since X_2 and $X_1 \cap X_2$ are both Stein and $X_1 \cap X_2$ is Runge in X_2 , it follows that $H^1(X_1 \cap X_2,\mathcal{F}) = 0$ and

$$\mathrm{Im}\Big(\Gamma(X_2,\mathcal{F}) \longrightarrow \Gamma(X_1 \cap X_2,\mathcal{F})\Big)$$

is dense in $\Gamma(X_1 \cap X_2, \mathcal{F})$. The proposition now follows from (5.4). Q.E.D.

(5.6) Proposition. *Suppose* $0 \leq \tilde{a} < a^* \leq a < b \leq b^* < \tilde{b}$ *in* $\mathbb{R}^N$, D *is an open neighborhood of* 0 *in* $\mathbb{C}^n$, *and* $\mathcal{F}$ *is a locally free sheaf on* $D \times G^N(\tilde{a},\tilde{b})$. *If* $N \geq 3$, *then there exists an open neighborhood* U *of* 0 *in* D *such that, for every connected Stein open neighborhood* W *of* 0 *in* U , $H^1(W \times G^N(a^*,b^*),\mathcal{F})$ *and* $H^1(W \times G^N(a,b),\mathcal{F})$ *are both Hausdorff and the restriction map* $H^1(W \times G^N(a^*,b^*),\mathcal{F}) \longrightarrow H^1(W \times G^N(a,b),\mathcal{F})$ *is a topological isomorphism.*

Proof. We can assume without loss of generality that $a^* < a$ and $b < b^*$. We need only show that there exists an open neighborhood U of 0 in D such that, for every connected Stein open neighborhood W of 0 in U , $H^1(W \times G^N(a,b),\mathcal{F})$ is Hausdorff.

Tale $a^* < a' < a$ and $b < b' < b^*$. Let $\mathfrak{U} = \{U_j\}_{j=1}^N$, where

$$U_j = \Delta^N(b') \cap \{|z_j| > a_j'\} .$$

By (5.3) there exists an open neighborhood U of 0 in D such that, for every connected Stein open neighborhood W of 0 in U , $H^1_{L^2_v}(W \times \mathfrak{U},\mathcal{F})$ is Hausdorff. We claim that this U satisfies the requirement.

Let W be a connected Stein open neighborhood of 0 in U . Consider the following commutative diagram of restriction maps

$$\begin{array}{ccc} H^1(W \times G(a^*,b^*),\mathcal{F}) & \xrightarrow{\theta} & H^1(W \times G(a,b),\mathcal{F}) \\ & \alpha \searrow \quad \nearrow & \\ & H^1_{L^2_v}(W \times \mathcal{U},\mathcal{F}) \,. & \end{array}$$

Let N^* (respectively N) be the topological closure of $\{0\}$ in $H^1(W \times G(a^*,b^*),\mathcal{F})$ (respectively $H^1(W \times G(a,b),\mathcal{F})$). By (5.5), $\theta(N^*) = N$. Since $H^1_{L^2_v}(W \times \mathcal{U},\mathcal{F})$ is Hausdorff, $\alpha(N^*) = 0$. It follows that $N = 0$. $H^1(W \times G^N(a,b),\mathcal{F})$ is Hausdorff.

Q.E.D.

(5.7) Theorem. Suppose $0 \leq a < b$ in $\mathbb{R}^N$, D is an open subset of $\mathbb{C}^n$, and $\mathcal{F}$ is a locally free sheaf on $D \times G^N(a,b)$. If $N \geq 3$, then $\mathcal{F}$ can be extended uniquely to a coherent analytic sheaf $\tilde{\mathcal{F}}$ on $D \times \Delta^N(b)$ satisfying $\tilde{\mathcal{F}}^{[n]} = \tilde{\mathcal{F}}$.

Proof. Fix $t \in D$. Take $a < a' < b' < b'' < b$ in $\mathbb{R}^N$. By (5.6) there exists a Stein open neighborhood W of 0 in D such that the restriction map

$$\sigma\colon H^1(W \times G^N(a',b''),\mathcal{F}) \longrightarrow H^1(W \times G^N(a',b'),\mathcal{F})$$

is an isomorphism. Take $x \in W \times G^N(b',b'')$. $|z_j(x)| > b_j'$ for some $1 \leq j \leq N$. Let $f = z_j - z_j(x)$. Consider the following exact sequence on $D \times G^N(a,b)$:

$$0 \longrightarrow \mathcal{F} \xrightarrow{\alpha} \mathcal{F} \longrightarrow \mathcal{F}/f\mathcal{F} \longrightarrow 0\,,$$

where α is defined by multiplication by f. We have the following commutative diagram with exact rows:

$$\Gamma(W \times G^N(a',b''),\mathcal{F}) \xrightarrow{\lambda} \Gamma(W \times G^N(a',b''),\mathcal{F}/f\mathcal{F})$$

$$\xrightarrow{\mu} H^1(W \times G^N(a',b''),\mathcal{F}) \xrightarrow{\nu} H^1(W \times G^N(a',b''),\mathcal{F})$$

$$\sigma \downarrow \qquad\qquad\qquad \tau \downarrow$$

$$H^1(W \times G^N(a',b'),\mathcal{F}) \xrightarrow{\theta} H^1(W \times G^N(a',b'),\mathcal{F}) .$$

Since f is nowhere zero on $W \times G^N(a',b')$, θ is an isomorphism. Hence $\mu = \sigma^{-1}\theta^{-1}\tau\nu\mu = 0$. λ is surjective. Since

$$(W \times G^N(a',b'')) \cap \operatorname{Supp}(\mathcal{F}/f\mathcal{F}) = (W \times \Delta^N(b'')) \cap \{z_j = z_j(x)\}$$

is Stein, the map

$$\Gamma(W \times G^N(a',b''),\mathcal{F}/f\mathcal{F}) \longrightarrow \mathcal{F}_x/\mathfrak{m}_x\mathcal{F}_x$$

is surjective, where $\mathfrak{m}_x$ is the maximum ideal of ${}_{n+N}\mathcal{O}_x$. Since λ is surjective, the map

$$\Gamma(W \times G^N(a',b''),\mathcal{F}) \longrightarrow \mathcal{F}_x/\mathfrak{m}_x\mathcal{F}_x$$

is surjective. By Nakayama's lemma, $\Gamma(W \times G^N(a',b''),\mathcal{F})$ generates $\mathcal{F}_x$. Since x is an arbitrary point of $W \times G^N(b',b'')$, $\Gamma(W \times G^N(a',b''),\mathcal{F})$ generates $\mathcal{F}|W \times G^N(b',b'')$. Take a relatively compact open neighborhood W^* of t in W and take $b' < a^* < b^* < b''$ in $\mathbb{R}^N$. Then there exists a sheaf-epimorphism $\varphi: {}_{n+N}\mathcal{O}^p \longrightarrow \mathcal{F}$ on $W^* \times G^N(a^*,b^*)$. Applying the above argument to $\operatorname{Ker} \varphi$ instead of $\mathcal{F}$, we obtain an open neighborhood $W^\#$ of t in W^* and obtain $a^* < a^\# < b^\# < b^*$ in $\mathbb{R}^N$ such that there exists an exact sequence of sheaf-homomorphisms

$${}_{n+N}\mathcal{O}^q \xrightarrow{\psi} {}_{n+N}\mathcal{O}^p \xrightarrow{\varphi} \mathcal{F} \longrightarrow 0$$

on $W^{\#} \times G^N(a^{\#}, b^{\#})$. Ψ can be extended to a sheaf-homomorphism $\tilde{\Psi}: {}_{n+N}\mathcal{O}^q \longrightarrow {}_{n+N}\mathcal{O}^p$ on $W^{\#} \times \Delta(b^{\#})$. Let $\mathcal{G}$ be the coherent analytic sheaf on $W^{\#} \times \Delta^N(b)$ which agrees with $\mathcal{F}$ on $W^{\#} \times G^N(a^{\#}, b)$ and agrees with Coker $\tilde{\Psi}$ on $W^{\#} \times \Delta^N(b^{\#})$. Let $\tilde{\mathcal{G}} = (\mathcal{G}/0_{[n+1]}\mathcal{G})^{[n]}$. By (3.16), $\tilde{\mathcal{G}}|W^{\#} \times G^N(a,b)$ is isomorphic to $\mathcal{F}|W^{\#} \times G^N(a,b)$. Since t is an arbitrary point of D , the theorem follows from (3.16). Q.E.D.

(5.8) Corollary. <u>Suppose</u> $0 \leq a < b$ <u>in</u> $\mathbb{R}^3$, D <u>is an open subset of</u> $\mathbb{C}^n$, <u>and</u> $\mathcal{F}$ <u>is a coherent analytic sheaf on</u> $D \times G^3(a,b)$ <u>such that</u> $\mathcal{F}^{[n+1]} = \mathcal{F}$. <u>Then</u> $\mathcal{F}$ <u>can be extended uniquely to a coherent analytic sheaf</u> $\tilde{\mathcal{F}}$ <u>on</u> $D \times \Delta^3(b)$ <u>satisfying</u> $\tilde{\mathcal{F}}^{[n]} = \tilde{\mathcal{F}}$.

<u>Proof</u>. Let $V = S_{n+2}(\mathcal{F})$. By (3.13), $\dim V \leq n$.

(a) Consider first the special case where $\dim(\{t\} \times G^3(a,b)) \cap V \leq 0$ for all $t \in D$. Fix $t \in D$. There exist $a \leq a' < b' \leq b$ in $\mathbb{R}^3$ and an open neighborhood U of t in D such that $(U \times G^3(a',b')) \cap V = \emptyset$. By (5.7), $\mathcal{F}|U \times G^3(a',b')$ extends to a coherent analytic sheaf $\mathcal{G}$ on $U \times \Delta^3(b')$ such that $\mathcal{G}^{[n]} = \mathcal{G}$. Let $\tilde{\mathcal{G}}$ be the sheaf on $U \times \Delta^3(b)$ which agrees with $\mathcal{F}$ on $U \times G^3(a',b)$ and agrees with $\mathcal{G}$ on $U \times \Delta^3(b')$. By (3.16), $\tilde{\mathcal{G}}$ extends $\mathcal{F}|U \times G^3(a,b)$. Since t is an arbitrary point of D , the special case follows from (3.16).

(b) For the general case, take an arbitrary relatively compact open subset D' of D . By (2.2) there exists an 3-dimensional plane T in $\mathbb{C}^n \times \mathbb{C}^3$ such that, for some open neighborhood R of D' in D we have

(i) $D' \times \Delta^3(b) \subset (R+T) \cap (\mathbb{C}^n \times \Delta^3(b)) \subset D \times \Delta^3(b)$,

(ii) $\dim(x+T) \cap V \leq 0$ for $x \in V$,

where $R+T = \{x+y \mid x \in R,\ y \in T\}$ and $x+T = \{x+y \mid y \in T\}$. By (a), $\mathcal{F} \mid (R+T) \cap (\mathbb{C}^n \times G^3(a,b))$ extends to a coherent analytic sheaf $\mathcal{G}$ on $(R+T) \cap (\mathbb{C}^n \times \Delta^3(b))$ such that $\mathcal{G}^{[n]} = \mathcal{G}$. Since D' is an arbitrary relatively compact open subset of D and since $\mathcal{G} \mid D' \times \Delta^3(b)$ extends $\mathcal{F} \mid D' \times G^3(a,b)$, the general case follows from (3.16). Q.E.D.

Appendix of Chapter 5

I. Topology of Sheaf Cohomology Groups.

(5.A.1) Proposition. Suppose X is a complex space (with countable topology), $\mathfrak{U}$ is a countable Stein covering of X, and $\mathcal{F}$ is a coherent analytic sheaf on X. Then for $p \geq 0$ the quotient topology of $H^p(X,\mathcal{F})$ induced from the Fréchet space $Z^p(\mathfrak{U},\mathcal{F})$ is independent of the covering $\mathfrak{U}$.

Proof. Let $\mathfrak{V}$ be another countable Stein covering of X. Let $\mathfrak{W} = \mathfrak{U} \cup \mathfrak{V}$. It suffices to show that the map

$$H^p(\mathfrak{W},\mathcal{F}) \longrightarrow H^p(\mathfrak{U},\mathcal{F})$$

defined by restriction is open. This follows from the fact that, in the following commutative diagram

$$\begin{array}{ccc} Z^p(\mathfrak{W},\mathcal{F}) & \xrightarrow{\alpha} & Z^p(\mathfrak{U},\mathcal{F}) \\ \downarrow & & \downarrow \beta \\ H^p(\mathfrak{W},\mathcal{F}) & \longrightarrow & H^p(\mathfrak{U},\mathcal{F}) \end{array},$$

α is a continuous epimorphism of Fréchet spaces and β is open. Q.E.D.

(5.A.2) Proposition. Suppose X is a complex space, $\mathcal{F}$ is a coherent analytic sheaf on X, and X_1, X_2 are open subsets of X such that $X = X_1 \cup X_2$ and $(X_1 - X_2)^- \cap (X_2 - X_1)^- = \emptyset$. Then the maps in the Mayer-Vietoris sequence

$$\cdots \longrightarrow H^{p-1}(X_1 \cap X_2, \mathcal{F}) \longrightarrow H^p(X, \mathcal{F}) \longrightarrow$$

$$H^p(X_1, \mathcal{F}) \oplus H^p(X_2, \mathcal{F}) \longrightarrow H^p(X_1 \cap X_2, \mathcal{F}) \longrightarrow \cdots$$

<u>are all continuous</u>.

<u>Proof</u>. Since

$$(X_1 - X_2)^- \cap (X_2 - X_1)^- = \emptyset ,$$

we can choose a countable Stein covering $\mathfrak{U}$ such that

(i) $\mathfrak{U}_j := \{U \subset \mathfrak{U} \mid U \subset X_j\}$ covers X_j $(j = 1,2)$

(ii) $\mathfrak{U}_{12} := \mathfrak{U}_1 \cap \mathfrak{U}_2$ covers $X_1 \cap X_2$,

(iii) $\mathfrak{U} = \mathfrak{U}_1 \cup \mathfrak{U}_2$,

(iv) if $U_j \in \mathfrak{U} - \mathfrak{U}_j$ $(j = 1,2)$, then $U_1 \cap U_2 = \emptyset$.

Let

$$\sigma_j{}^p : C^p(\mathfrak{U}, \mathcal{F}) \longrightarrow C^p(\mathfrak{U}_j, \mathcal{F})$$

and

$$\tau_j{}^p : C^p(\mathfrak{U}_j, \mathcal{F}) \longrightarrow C^p(\mathfrak{U}_{12}, \mathcal{F})$$

be restriction maps $(j = 1,2)$. Consider the following diagram:

$$(*) \qquad \begin{array}{ccccccccc}
 & & \vdots & & \vdots & & \vdots & & \\
 & & \downarrow & & \downarrow & & \downarrow & & \\
0 & \longrightarrow & C^{p-1}(\mathfrak{U}, \mathcal{F}) & \xrightarrow{\alpha_{p-1}} & C^{p-1}(\mathfrak{U}_1, \mathcal{F}) \oplus C^{p-1}(\mathfrak{U}_2, \mathcal{F}) & \xrightarrow{\beta_{p-1}} & C^{p-1}(\mathfrak{U}_{12}, \mathcal{F}) & \longrightarrow & 0 \\
 & & \downarrow \delta & & \downarrow \delta_1 \oplus \delta_2 & & \downarrow \delta_{12} & & \\
0 & \longrightarrow & C^p(\mathfrak{U}, \mathcal{F}) & \xrightarrow{\alpha_p} & C^p(\mathfrak{U}_1, \mathcal{F}) \oplus C^p(\mathfrak{U}_2, \mathcal{F}) & \xrightarrow{\beta_p} & C^p(\mathfrak{U}_{12}, \mathcal{F}) & \longrightarrow & 0 \\
 & & \downarrow & & \downarrow & & \downarrow & & \\
 & & \vdots & & \vdots & & \vdots & &
\end{array}$$

where

$$\alpha_p(a) = \sigma_1{}^p(a) \oplus \sigma_2{}^p(a) ,$$

$$\beta_p(a_1 \oplus a_2) = \tau_1{}^p(a_1) - \tau_2{}^p(a_2) ,$$

and δ, δ_1, δ_2, δ_{12} are coboundary maps. It follows from (i), (ii), (iii), (iv) that the rows of (*) are exact. The sequence

$$\ldots \longrightarrow H^{p-1}(\mathfrak{U}_{12},\mathcal{F}) \xrightarrow{\theta} H^p(\mathfrak{U},\mathcal{F}) \longrightarrow$$

$$H^p(\mathfrak{U}_1,\mathcal{F}) \oplus H^p(\mathfrak{U}_2,\mathcal{F}) \longrightarrow H^p(\mathfrak{U}_{12},\mathcal{F}) \longrightarrow \ldots$$

defined by (*) is isomorphic to the Mayer-Vietoris sequence in the statement of the proposition. To finish the proof, it suffices to show that θ is continuous when $H^{p-1}(\mathfrak{U}_{12},\mathcal{F})$ and $H^p(\mathfrak{U},\mathcal{F})$ are given respectively the quotient topology induced from the Fréchet spaces $Z^{p-1}(\mathfrak{U}_{12},\mathcal{F})$ and $Z^p(\mathfrak{U},\mathcal{F})$. Consider the following diagram:

$$\begin{array}{ccc}
\beta_{p-1}^{-1}\left(Z^{p-1}(\mathfrak{U}_{12},\mathcal{F})\right) & \xrightarrow{(\delta_1\oplus\delta_2)'} & (\delta_1\oplus\delta_2)\left(\beta_{p-1}^{-1}\left(Z^{p-1}(\mathfrak{U}_{12},\mathcal{F})\right)\right) \\
 & & \uparrow{\scriptstyle \alpha_p'} \\
\downarrow{\scriptstyle \beta_{p-1}'} & & \alpha_p^{-1}((\delta_1\oplus\delta_2)(\beta_{p-1}^{-1}(Z^{p-1}(\mathfrak{U}_{12},\mathcal{F})))) \\
 & & \downarrow{\scriptstyle i} \\
Z^{p-1}(\mathfrak{U}_{12},\mathcal{F}) \xrightarrow{\xi} H^{p-1}(\mathfrak{U}_{12},\mathcal{F}) & \xrightarrow{\theta} & H^p(\mathfrak{U},\mathcal{F}) \xleftarrow{\eta} Z^p(\mathfrak{U},\mathcal{F}) ,
\end{array}$$

where ξ, η are the natural quotient maps, and α_p', β_{p-1}', $(\delta_1\oplus\delta_2)'$ are respectively induced by α_p, β_{p-1}. $\delta_1\oplus\delta_2$. From the definition of θ , it follows that, if

$$u \in \beta_{p-1}^{-1}\left(Z^{p-1}(\mathfrak{U}_{12}, \mathcal{F})\right)$$

and

$$v \in \alpha_p^{-1}\left((\delta_1 \oplus \delta_2)\left(\beta_{p-1}^{-1} Z^{p-1}(\mathfrak{U}_{12}, \mathcal{F})\right)\right) ,$$

such that

$$(\delta_1 \oplus \delta_2)'(u) = \alpha_p'(v) ,$$

then

$$(\theta \xi \beta_{p-1}')(u) = (\eta i)(v) .$$

Since the rows of (*) are exact, α_p' and β_{p-1}' are both surjective. They are both open, because $\operatorname{Im} \alpha_p$ and $\operatorname{Im} \beta_{p-1}$ are closed. Since ξ is open by definition of the topology of $H^{p-1}(\mathfrak{U}_{12}, \mathcal{F})$, if G is an open subset of $H^p(\mathfrak{U}, \mathcal{F})$, then

$$\theta^{-1}(G) = \xi \beta_{p-1}'\left((\delta_1 \oplus \delta_2)'\right)^{-1} \alpha_p' i^{-1} \eta^{-1}(G)$$

is open in $H^{p-1}(\mathfrak{U}_{12}, \mathcal{F})$. Hence θ is continuous. Q.E.D.

II. Triviality of Holomorphic Vector Bundles.

We will prove, by using Cartan's lemma, the following special case of Grauert's theorem: every holomorphic vector bundle on a polydisc or the product of an annulus and a polydisc is trivial. First, we introduce the following notations: for an open subset D of $\mathbb{C}^n$,

$B(D)$ = the Banach space of all uniformly bounded holomorphic functions on D ,

$\mathcal{M}_r$ = the sheaf of germs of $r \times r$ matrices of holomorphic functions on $\mathbb{C}^n$,

$B(D,\mathcal{M}_r)$ = the Banach algebra of all $r \times r$ matrices of uniformly bounded holomorphic functions on D ,

$\mathcal{G}\ell_r$ = the sheaf of germs of nonsingular $r \times r$ matrices of holomorphic functions on $\mathbb{C}^n$,

$B(D,\mathcal{G}\ell_r)$ = the group of all $M \in B(D,\mathcal{M}_r)$ such that $M^{-1} \in B(D,\mathcal{M}_r)$,

$\Gamma_0(D,\mathcal{G}\ell_r)$ = the pathwise connected identity component of $\Gamma(D,\mathcal{G}\ell_r)$ furnished with the topology of uniform convergence on compact subsets.

(5.A.3) Lemma. Suppose G is a bounded domain in $\mathbb{C}$ and D is an open subset of $\mathbb{C}^n$. Let E be the space of all uniformly bounded C^∞ functions on $D \times G$. Then there exists a linear map $\varphi: E \longrightarrow E$ such that $\frac{\partial}{\partial \bar{z}} \varphi(f) = f$ and $\|\varphi(f)\|_{D \times G} \leq C\|f\|_{D \times G}$ for $f \in E$, where z is the coordinate of G , $\|\cdot\|_{D \times G}$ is the sup norm on $D \times G$, and C is a constant independent of f .

Proof. For $f \in E$ define

$$\varphi(f)(t,z) = \frac{1}{2\pi i} \int_{\zeta \in G} \frac{f(t,\zeta)d\zeta \wedge d\bar{\zeta}}{\zeta - z} .$$

We claim that φ satisfies the requirement. Clearly

$$\|\varphi(f)\|_{D \times G} \leq C\|f\|_{D \times G} ,$$

where

$$C = \frac{1}{2\pi} \int_{\zeta \in \Omega} \frac{|d\zeta \wedge d\bar{\zeta}|}{|\zeta|}$$

and

$$\Omega = \{z_1 - z_2 | z_1, z_2 \in G\} .$$

To verify

$$\frac{\partial}{\partial \bar{z}} \varphi(f) = f ,$$

take an arbitrary compact subset K of G. Choose an open neighborhood L of K in G such that $L \subset\subset G$ and the boundary of L is piecewise smooth. By the standard technique of applying Stokes' theorem to the form $d\left(\frac{f d\zeta}{\zeta - z}\right)$ on $L - \{z\}$, we conclude that

$$\frac{\partial}{\partial \bar{z}}\left(\frac{1}{2\pi i} \int_L \frac{f(t,\zeta) d\zeta \wedge d\bar{\zeta}}{\zeta - z}\right) = f(t,z) \quad \text{for} \quad t \in D \quad \text{and} \quad z \in L .$$

Hence

$$\left(\frac{\partial}{\partial \bar{z}}(\varphi f)\right)(t,z)$$

$$= \frac{\partial}{\partial \bar{z}} \frac{1}{2\pi i} \int_L \frac{f(t,\zeta) d\zeta \wedge d\bar{\zeta}}{\zeta - z} + \frac{\partial}{\partial \bar{z}} \frac{1}{2\pi i} \int_{G-L} \frac{f(t,\zeta) d\zeta \wedge d\bar{\zeta}}{\zeta - z}$$

$$= f(t,z)$$

for $t \in D$ and $z \in L$, because

$$\frac{1}{2\pi i} \int_{G-L} \frac{f(t,\zeta) d\zeta \wedge d\bar{\zeta}}{\zeta - z}$$

is holomorphic in z for $t \in D$ and $z \in L$. The lemma follows from the arbitrariness of K. Q.E.D.

(5.A.4) Lemma. *Suppose* G *is a bounded domain in* $\mathbb{C}$ *and* $\tilde{G}_j$ *is an open subset of* $\mathbb{C}$ $(j = 1,2)$ *such that*

$G \subset\subset \tilde{G}_1 \cup \tilde{G}_2$. Let $G_j = G \cap \tilde{G}_j$ $(j = 1,2)$ and $G_{12} = G_1 \cap G_2$. Suppose D is an open subset of $\mathbb{C}^n$. Then there exists a continuous linear map φ_j: $B(D \times G_{12}) \longrightarrow B(D \times G_j)$ $(j = 1,2)$ such that $\varphi_1 f + \varphi_2 f = f$ on $D \times G_{12}$ for $f \in B(D \times G_{12})$.

Proof. Choose a C^∞ function ρ on $\mathbb{C}$ such that $0 \leq \rho \leq 1$, $\operatorname{Supp} \rho \subset\subset \tilde{G}_1$, and $\rho \equiv 1$ in an open neighborhood of $\bar{G}-\tilde{G}_2$. Since $\operatorname{Supp} \frac{\partial \rho}{\partial \bar{z}}$ is a compact subset of $\tilde{G}_1$ and is disjoint from $\bar{G}-\tilde{G}_2$, we have

$$G \cap \operatorname{Supp} \frac{\partial \rho}{\partial \bar{z}} \subset G_{12}.$$

By (5.A.3) there exists a linear map Ψ from the space E of all uniformly bounded C^∞ functions on $D \times G$ to itself such that $\frac{\partial}{\partial \bar{z}} \Psi(g) = g$ for $g \in E$. For $f \in B(D \times G_{12})$ define

$$\varphi_1(f) = (1-(\rho \circ \pi)) f + \Psi\left(\left(\frac{\partial \rho}{\partial \bar{z}} \circ \pi\right) f\right),$$

$$\varphi_2(f) = (\rho \circ \pi) f - \Psi\left(\left(\frac{\partial \rho}{\partial \bar{z}} \circ \pi\right) f\right),$$

where $\pi: D \times G \longrightarrow G$ is the natural projection. Then φ_1, φ_2 satisfy the requirements. Q.E.D.

(5.A.5) Lemma. Suppose G is a bounded domain in $\mathbb{C}$ and $\tilde{G}_j$ is an open subset of $\mathbb{C}$ $(j = 1,2)$ such that $G \subset\subset \tilde{G}_1 \cup \tilde{G}_2$. Let $G_j = G \cap \tilde{G}_j$ $(j = 1,2)$ and $G_{12} = G_1 \cap G_2$. Suppose D is an open subset of $\mathbb{C}^n$. Then for some $0 < \eta < 1$ there exists a continuous map θ_j from $A := \{M \in \Gamma(D \times G_{12}, \mathfrak{M}_r) \mid \|M\|_{D \times G_{12}} < \eta\}$ to $B(D \times G_j, \mathcal{GL}_r)$ $(j = 1,2)$ such that $I+M = \theta_1(M)^{-1}\theta_2(M)$ on $D \times G_{12}$, where

$\|M\|_{D \times G_{12}}$ _denotes the maximum of the sup norms on_ $D \times G_{12}$ _of the entries of_ M .

Proof. By (5.A.4), there exists a continuous linear map

$$\Phi_j : B(D \times G_{12}, \mathfrak{M}_r) \longrightarrow B(D \times G_j, \mathfrak{M}_r)$$

$(j = 1,2)$ such that

$$\Phi_1(M) + \Phi_2(M) = M$$

on $D \times G_{12}$ for $M \in B(D \times G_{12}, \mathfrak{M}_r)$. There exists $C \geq 1$ such that, for $j = 1,2$,

$$(*) \qquad \|\Phi_j(M)\|_{D \times G_j} \leq C\|M\|_{D \times G_{12}}$$

for $M \in B(D \times G_{12}, \mathfrak{M}_r)$.

Take $0 < \eta < \frac{1}{4r^2C}$. We claim that η satisfies the requirement. Observe that,

$$(\dagger) \qquad \begin{cases} \text{if } P \text{ is an } r \times r \text{ scalar matrix with } \|I-P\| < C\eta , \\ \text{then } P^{-1} \text{ exists and } \|P^{-1}\| \leq 2 , \end{cases}$$

because

$$P^{-1} = \big(I-(I-P)\big)^{-1} = \sum_{\nu=0}^{\infty} (I-P)^{\nu} .$$

Define, by induction on ν , maps

$$\sigma_j^{\nu} : A \longrightarrow B(D \times G_j, \mathfrak{M}_r) ,$$

$\sigma^{\nu} : A \longrightarrow A \quad (j = 1,2;\ \nu \geq 0)$ as follows:

(i) $\sigma^0(M) = M$,

(ii) $\sigma_j^{\nu}(M) = (-1)^j(\Phi_j \circ \sigma^{\nu})(M)$,

(iii) $\sigma^{\nu+1}(M) = \sigma_1^{\nu}(M)\sigma^{\nu}(M)(1+\sigma_2^{\nu}(M))^{-1}$.

We claim that $\|\sigma^{\nu}(M)\|_{D \times G_{12}} \leq \frac{\eta}{2^{\nu}}$ for $M \in A$ so that, by (*),

$$\|\sigma_j^{\nu}(M)\|_{D \times G_j} \leq \frac{C\eta}{2^{\nu}}$$

$(j = 1,2)$ and, by (†), $(I+\sigma_2^{\nu}(M))^{-1}$ exists, making the definition of $\sigma^{\nu+1}$ meaningful, and

$$\|(I+\sigma_2^{\nu}(M))^{-1}\|_{D \times G_2} \leq 2 .$$

The claim follows from induction on ν and the following inequality coming from (iii):

$$\|\sigma^{\nu+1}(M)\|_{D\times G_{12}} \leq r^2 \|\sigma_1^{\nu}(M)\|_{G\times D_1} \|\sigma^{\nu}(M)\|_{D\times G_{12}} \|(I+\sigma_2^{\nu}(M))^{-1}\|_{D\times G_2}$$

$$\leq r^2 \cdot \frac{C\eta}{2^{\nu}} \cdot \frac{\eta}{2^{\nu}} \cdot 2 \leq \frac{\eta}{2^{2\nu+1}} \leq \frac{\eta}{2^{\nu+1}} .$$

It follows from (iii) and (ii) that

$$\begin{aligned}(I+\sigma^{\nu+1}(M))(I+\sigma_2^{\nu}(M)) &= I + \sigma_2^{\nu}(M) + \sigma^{\nu+1}(M)(I+\sigma_2^{\nu}(M)) \\ &= I + \sigma_2^{\nu}(M) + \sigma_1^{\nu}(M)\sigma^{\nu}(M) \\ &= I + \sigma_1^{\nu}(M) + \sigma^{\nu}(M) + \sigma_1^{\nu}(M)\sigma^{\nu}(M) \\ &= (I+\sigma_1^{\nu}(M))(I+\sigma^{\nu}(M)) .\end{aligned}$$

Hence

$$I + \sigma^{\nu+1}(M) = (I+\sigma_1{}^{\nu}(M))(I+\sigma^{\nu}(M))(I+\sigma_2{}^{\nu}(M))^{-1} .$$

Define

$$\tau_j{}^{\nu}: A \longrightarrow B(D \times G_j, \mathcal{Gl}_r)$$

$(j = 1,2;\ \nu \geq 0)$ as follows:

$$\tau_j{}^{\nu}(M) = (I+\sigma_j{}^{\nu}(M))(I+\sigma_j^{\nu-1}(M)) \ldots (I+\sigma_j{}^{0}(M)) .$$

Then

$$I + \sigma^{\nu+1}(M) = \tau_1{}^{\nu}(M)(I+\sigma^{0}(M))(\tau_2{}^{\nu}(M))^{-1}.$$

Define

$$\theta_j: A \longrightarrow B(D \times G_j, \mathcal{Gl}_r)$$

by

$$\theta_j(M) = \lim_{\nu \to \infty} \tau_j{}^{\nu}(M)$$

$(j = 1,2)$. Since

$$\|\sigma_j{}^{\nu}(M)\|_{D \times G_j} \leq \frac{C\eta}{2^{\nu}}$$

for $0 \leq \nu < \infty$, θ_j is well-defined $(j = 1,2)$. Moreover,

$$I + M = \theta_1(M)^{-1}\theta_2(M)$$

on $D \times G_{12}$ for $M \in A$. Q.E.D.

(5.A.6) __Lemma.__ __Suppose__ $D \subset \tilde{D}$ __are Stein open subsets of__ $\mathbb{C}^n$ __and__ D __is Runge in__ $\tilde{D}$. __Suppose__ K __is a compact subset of__ D __and__ $\epsilon > 0$. __Suppose__ $t \longmapsto f_t$ __is a continuous map from__ $[0,1]$ __to__ $\Gamma(D, \mathcal{Gl}_r)$ __with__ $f_0 = I$. __Then there exists a__ continuous map $t \longmapsto g_t$ from $[0,1]$ to $\Gamma(\tilde{D}, \mathcal{G}\ell_r)$ such that $\|I-g_t^{-1}f_t\|_K < \epsilon$ for $0 \leq t \leq 1$, and $g_0 = I$.

Proof. We can assume without loss of generality that K is holomorphically convex in D. Take a relatively compact open neighborhood L of K in D. First we show that,

(*) for every continuous map $t \longmapsto \xi_t$ from $[0,1]$ to $\Gamma(L, \mathfrak{m}_r)$ with $\xi_0 = 0$ and every $\gamma > 0$ there exists a continuous map $t \longmapsto \eta_t$ from $[0,1]$ to $\Gamma(\tilde{D}, \mathfrak{m}_r)$ with $\eta_0 = 0$ such that $\|\xi_t - \eta_t\|_K < \gamma$ for $0 \leq t \leq 1$.

Take $\delta > 0$ and

$$0 = t_0 < t_1 < \dots < t_k = 1$$

with

$$t_{j+1} - t_j < \delta \quad \text{for } 0 \leq j < k.$$

Since D is Runge in $\tilde{D}$ and K is holomorphically convex in D, there exists

$$\zeta_j \in \Gamma(\tilde{D}, \mathfrak{m}_r)$$

such that

$$\|\xi_{t_j} - \zeta_j\|_K < \delta$$

$(1 \leq j \leq k)$. Let $\zeta_0 = 0$. Define η_t by

$$\eta_t = \frac{(t-t_{j-1})\zeta_j + (t_j - t)\zeta_{j-1}}{t_j - t_{j-1}} \quad \text{for } t_{j-1} \leq t \leq t_j.$$

When δ is small enough, $\|\xi_t - \eta_t\|_K < \gamma$ for $0 \leq t \leq 1$. (*) is proved.

To finish the proof of the lemma, take $\gamma > 0$ and take

$$0 = t_0 < t_1 < \ldots < t_k = 1$$

with

$$t_{j+1} - t_j < \gamma \quad (0 \leq j < k) .$$

When γ is small enough,

$$\left\| I - f_t^{-1} f_{t_j} \right\|_L < \frac{1}{2r}$$

for $t_j \leq t \leq t_{j+1}$ $(0 \leq j < k)$. Define a continuous map $t \longmapsto h_{j,t}$ from $[t_j, t_{j+1}]$ to $\Gamma(L, \mathcal{G}\ell_r)$ by

$$h_{j,t} = - \sum_{\nu=0}^{\infty} \frac{\left(I - f_t^{-1} f_{t_j}\right)^{\nu+1}}{\nu+1}$$

$(0 \leq j < k)$. Then $\exp h_{j,t} = f_t^{-1} f_{t_j}$ for $t_j \leq t \leq t_{j+1}$ $(0 \leq j < k)$. By (*) there exists a continuous map $t \longmapsto \tilde{h}_{j,t}$ from $[t_j, t_{j+1}]$ to $\Gamma(\tilde{D}, \mathcal{G}\ell_r)$ such that

$$\left\| \tilde{h}_{j,t} - h_{j,t} \right\|_K < \gamma$$

for $t_j \leq t \leq t_{j+1}$ and $h_{j,t_j} = 0$ $(0 \leq j < k)$. Define g_t by

$$g_t = (\exp h_{j,t})(\exp h_{j-1,t_j}) \ldots (\exp h_{0,t_1})$$

for $t_j \leq t \leq t_{j+1}$ $(0 \leq j < k)$. When γ is small enough $\left\| I - g_t^{-1} f_t \right\|_K < \epsilon$ for $0 \leq t \leq 1$. Q.E.D.

(5.A.7) Proposition. Every holomorphic vector bundle on an open polydisc is trivial.

Proof. It is equivalent to prove that every holomorphic vector bundle B on an open polysquare R is trivial.

First, we show that, for every open polysquare S relatively compact in R, the restriction of B to S is trivial. Let

$$R = R_1 \times \ldots \times R_n$$

and

$$S = S_1 \times \ldots \times S_n ,$$

where $R_j, S_j \subset \mathbb{C}$ $(1 \leq j \leq n)$. By dividing up R into a grid of small polysquares, it suffices to prove the triviality of $B|S$ with the following additional assumption: there exist open squares $R_1^{(\nu)} \subset \mathbb{C}$ $(\nu = 1,2)$ such that $R_1 = R_1^{(1)} \cup R_1^{(2)}$ and $B|R^{(\nu)}$ is trivial, where

$$R^{(\nu)} = R_1^{(\nu)} \times R_2 \times \ldots \times R_n$$

$(\nu = 1,2)$. B is defined by some

$$M \in \Gamma(R^{(1)} \cap R^{(2)}, \mathcal{GL}_r) .$$

Choose open squares $S_1^{(\nu)} \subset \mathbb{C}$ $(\nu = 1,2)$ such that $S_1 = S_1^{(1)} \cup S_1^{(2)}$ and $S_1^{(\nu)} \subset\subset R_1^{(\nu)}$ $(\nu = 1,2)$. Let

$$S^{(\nu)} = S_1^{(\nu)} \times S_2 \times \ldots \times S_n$$

$(\nu = 1,2)$. Since $R^{(1)} \cap R^{(2)}$ is starlike,

$$\Gamma_0(R^{(1)} \cap R^{(2)}, \mathcal{GL}_r) = \Gamma(R^{(1)} \cap R^{(2)}, \mathcal{GL}_r) .$$

By (5.A.6), for every $\epsilon > 0$ there exists $\tilde{M} \in \Gamma(R, \mathcal{Gl}_r)$ such that

$$\|I-\tilde{M}^{-1}M\|_{S^{(1)} \cap S^{(2)}} < \epsilon .$$

By (5.A.5), when ϵ is small enough, there exist

$$M_j \in B(S^{(j)}, \mathcal{Gl}_r)$$

$(j = 1,2)$ such that $\tilde{M}^{-1}M = M_1^{-1}M_2$ on $S^{(1)} \cap S^{(2)}$. Hence $M = (M_1\tilde{M}^{-1})^{-1}M_2$ on $S^{(1)} \cap S^{(2)}$ and $B|S$ is trivial.

Take a sequence of open polysquares U_α in R $(1 \leq \alpha < \infty)$ such that $U_\alpha \subset\subset U_{\alpha+1}$ and $R = \bigcup_{\alpha=1}^{\infty} U_\alpha$. Since $B|U_\alpha$ is trivial for each α, B is defined by the covering $\{U_\alpha\}_{\alpha=1}^{\infty}$ and transition functions $\{M_{\alpha\beta}\}_{\alpha=1}^{\infty}$, where $M_{\alpha\beta} \in \Gamma(U_\alpha \cap U_\beta, \mathcal{Gl}_r)$.

We are going to construct, by induction on ν, $P_\alpha^{(\nu)} \in \Gamma(U_\alpha, \mathcal{Gl}_r)$ $(1 \leq \alpha \leq \nu < \infty)$ such that

(i) $(P_\alpha^{(\nu)})^{-1}P_\beta^{(\nu)} = M_{\alpha\beta}$ on $U_{\alpha\beta}$ for $1 \leq \alpha, \beta \leq \nu$,

(ii) $\|I-P_\alpha^{(\nu+1)}(P_\alpha^{(\nu)})^{-1}\|_{U_\alpha} < \frac{1}{2^\nu}$ for $1 \leq \alpha \leq \nu-1$.

Define $P_1^{(1)} = I$. Suppose we have constructed $P_\alpha^{(\nu)}$ for $1 \leq \alpha \leq \nu$. By (5.A.6) there exists $Q \in \Gamma(R, \mathcal{Gl}_r)$ such that $\|I-Q^{-1}M_{\nu+1,\nu}(P_\nu^{(\nu)})^{-1}\|_{U_{\nu-1}} < \frac{1}{2^\nu}$. Define $P_\alpha^{(\nu+1)} = Q^{-1}M_{\nu+1,\alpha}$ for $1 \leq \alpha \leq \nu+1$. Then

$$M_{\alpha\beta} = M_{\nu+1,\alpha}{}^{-1}M_{\nu+1,\beta} = \left(P_\alpha^{(\nu+1)}\right)^{-1}P_\beta^{(\nu+1)}$$

on $U_{\alpha\beta}$ for $1 \leq \alpha, \beta \leq \nu+1$. For $1 \leq \alpha \leq \nu$, since

$$\left(P_{\nu}^{(\nu)}\right)^{-1} P_{\alpha}^{(\nu)} = M_{\nu\alpha} = M_{\nu+1,\nu}^{-1} M_{\nu+1,\alpha}$$

on U_{α}, it follows that

$$M_{\nu+1,\alpha}\left(P_{\alpha}^{(\nu)}\right)^{-1} = M_{\nu+1,\nu}\left(P_{\nu}^{(\nu)}\right)^{-1}$$

on U_{α}. Hence condition (ii) is verified. The construction is complete.

For $1 \leq \alpha < \infty$ define $P_{\alpha} = \lim_{\nu \to \infty} P_{\alpha}^{(\nu)}$. $P_{\alpha} \in \Gamma(U_{\alpha}, \mathcal{G}\ell_r)$, because

$$P_{\alpha} = \lim_{\nu \to \infty} (I+Q_{\alpha}^{(\nu)})(I+Q_{\alpha}^{(\nu-1)}) \ldots (I+Q_{\alpha}^{(\alpha+1)}) P_{\alpha}^{(\alpha)}$$

where

$$Q_{\alpha}^{(\nu)} = P_{\alpha}^{(\nu)}\left(P_{\alpha}^{(\nu-1)}\right)^{-1} - I$$

and $\left\| Q_{\alpha}^{(\nu)} \right\|_{U_{\alpha}} < \frac{1}{2^{\nu}}$ for $\alpha < \nu < \infty$. Since $M_{\alpha\beta} = P_{\alpha}^{-1} P_{\beta}$ on $U_{\alpha\beta}$ for $1 \leq \alpha, \beta \leq \nu$, it follows that B is trivial on R. Q.E.D.

(5.A.8) <u>Proposition</u>. <u>Every holomorphic vector bundle on the product of an open annulus and an open polydisc is trivial.</u>

<u>Proof</u>. Take $0 < \eta < 1$ and suppose B is a holomorphic vector bundle on $G^1(\eta,1) \times \Delta^n$. We have to show that B is trivial. For $1 \leq \alpha < \infty$, choose $\eta < \eta_{\alpha} < r_{\alpha} < 1$ such that $\eta_{\alpha+1} < \eta_{\alpha}$, $r_{\alpha} < r_{\alpha+1}$, $\eta_{\alpha} \to \eta$, and $r_{\alpha} \to 1$ as $\alpha \to \infty$. Define

$$U_1 = (\overset{1}{G}(\eta,1) \times \Delta^n) \cap \{\mathrm{Re}\ w < \eta\} ,$$
$$U_2 = (\overset{1}{G}(\eta,1) \times \Delta^n) \cap \{\mathrm{Re}\ w > -\eta\} ,$$
$$U_1^{\alpha} = (\overset{1}{G}(\eta_\alpha,r_\alpha) \times \Delta^n(r_\alpha)) \cap \{\mathrm{Re}\ w < \eta\} ,$$
$$U_2^{\alpha} = (\overset{1}{G}(\eta_\alpha,r_\alpha) \times \Delta^n(r_\alpha)) \cap \{\mathrm{Re}\ w > -\eta\} ,$$

where w is the coordinate of $\overset{1}{G}(\eta,1)$. Let $U_{12} = U_1 \cap U_2$, $U = U_1 \cup U_2$, $U_{12}^{\alpha} = U_1^{\alpha} \cap U_2^{\alpha}$, and $U^{\alpha} = U_1^{\alpha} \cup U_2^{\alpha}$.

By (5.A.7), $B|U_j$ is trivial $(j = 1,2)$. Hence B is represented by some $M \in \Gamma(U_{12},\mathcal{Gl}_r)$. Since U_{12} is biholomorphic to the disjoint union of two open polydiscs,

$$\Gamma(U_{12},\mathcal{Gl}_r) = \Gamma_0(U_{12},\mathcal{Gl}_r) .$$

There exists a continuous map $t \longrightarrow M_t$ from $[0,1]$ to $\Gamma(U_{12},\mathcal{Gl}_r)$ such that $M_0 = I$ and $M_1 = M$.

Since U_{12} is Runge in U , by (5.A.6) for every $1 \le \alpha < \infty$ and every $\epsilon_\alpha > 0$ there exists a continuous map $t \longrightarrow Q_t^{(\alpha)}$ from $[0,1]$ to $\Gamma(U_, \mathcal{Gl}_r)$ such that $Q_0^{(\alpha)} = I$ and

$$\left\| I-(Q_t^{(\alpha)})^{-1}M_t \right\|_{U_{12}^{\alpha}} < \epsilon_\alpha$$

for $0 \le t \le 1$.

By (5.A.5), when ϵ_α is small enough, there exists a continuous map $t \longrightarrow \tilde{P}_{j,t}^{(\alpha)}$ from $[0,1]$ to $\Gamma(U_j^{\alpha},\mathcal{Gl}_r)$ $(j = 1,2)$ such that

$$(Q_t^{(\alpha)})^{-1}M_t = (\tilde{P}_{1,t}^{(\alpha)})^{-1}\tilde{P}_{2,t}^{(\alpha)}$$

on U_{12}^{α} $(1 \le \alpha < \infty)$. Let

$$P^{(\alpha)}_{1,t} = \left(\tilde{P}^{(\alpha)}_{1,0}\right)^{-1}\left(Q^{(\alpha)}_t\right)^{-1} \tilde{P}^{(\alpha)}_{1,t}$$

and $P^{(\alpha)}_{2,t} = \tilde{P}^{(\alpha)}_{2,t}\, \tilde{P}^{(\alpha)\,-1}_{2,0}$. Then $M_t = \left(P^{(\alpha)}_{1,t}\right)^{-1} P^{(\alpha)}_{2,t}$ on U^{α}_{12} and $P^{(\alpha)}_{j,0} = I$ $(j = 1,2)$.

Since

$$\left(P^{(\alpha)}_{1,t}\right)^{-1} P^{(\alpha)}_{2,t} = M_t = \left(P^{(\alpha+1)}_{1,t}\right)^{-1} P^{(\alpha+1)}_{2,t}$$

on U^{α}_{12} , it follows that $P^{(\alpha+1)}_{1,t}\left(P^{(\alpha)}_{1,t}\right)^{-1} = P^{(\alpha+1)}_{2,t}\left(P^{(\alpha)}_{2,t}\right)^{-1}$ on U^{α}_{12} $(0 \leq t \leq 1,\ 1 \leq \alpha < \infty)$. Define $R^{(\alpha)}_t \in \Gamma(U^{\alpha}, \mathcal{Gl}_r)$ by

$$R^{(\alpha)}_t = \begin{cases} P^{(\alpha+1)}_{1,t}\left(P^{(\alpha)}_{1,t}\right)^{-1} & \text{on } U_1{}^{\alpha} \\ P^{(\alpha+1)}_{2,t}\left(P^{(\alpha)}_{2,t}\right)^{-1} & \text{on } U_2{}^{\alpha} . \end{cases}$$

Let $R^{(\alpha)} = R^{(\alpha)}_1$. Since $R^{(\alpha)}_0 = I$, $R^{(\alpha)} \in \Gamma_0(U^{\alpha}, \mathcal{Gl}_r)$.

By (5.A.6), we can construct, by induction on α , $S^{(\alpha)} \in \Gamma_0(U, \mathcal{Gl}_r)$ $(1 \leq \alpha < \infty)$ such that $S^{(1)} = I$ and

$$\left\| I - (S^{(\alpha+1)})^{-1} R^{(\alpha)} S^{(\alpha)} \right\|_{U_{\alpha-1}} < \frac{1}{2^{\alpha}} ,$$

because U_{α} is Runge in U . Define $T^{(\alpha)}_j \in \Gamma(U_j{}^{\alpha}, \mathcal{Gl}_r)$ by $T^{(\alpha)}_j = (S^{(\alpha)})^{-1} P^{(\alpha)}_{j,1}$ $(j = 1,2;\ 1 \leq \alpha < \infty)$. Then $\left(T^{(\alpha)}_1\right)^{-1} T^{(\alpha)}_2 = M$ on U^{α}_{12} . Moreover,

$$\left\| I - T^{(\alpha+1)}_j \left(T^{(\alpha)}_j\right)^{-1} \right\|_{U_{\alpha-1}} < \frac{1}{2^{\alpha}}$$

$(j = 1,2)$, because

$$\begin{aligned} T^{(\alpha+1)}_j \left(T^{(\alpha)}_j\right)^{-1} &= (S^{(\alpha+1)})^{-1} P^{(\alpha+1)}_{j,1} P^{(\alpha)}_{j,1} S^{(\alpha)} \\ &= (S^{(\alpha+1)})^{-1} R^{(\alpha)} S^{(\alpha)} . \end{aligned}$$

Let $T_j \in \Gamma(U_j, \mathcal{G}\ell_r)$ be defined by $T_j = \lim_{\alpha \to \infty} T_j^{(\alpha)}$ $(j = 1,2)$. T_j is well-defined, because, for $\alpha < \beta$, on U_j^{α},

$$T_j^{(\beta)} = \left(I+Y_j^{(\beta)}\right)\left(I+Y_j^{(\beta-1)}\right) \cdots \left(I+Y_j^{(\alpha+1)}\right) T_j^{(\alpha)} ,$$

where

$$Y_j^{(\gamma)} = T_j^{(\gamma)}\left(T_j^{(\gamma-1)}\right)^{-1} - I$$

and

$$\left\| Y_j^{(\gamma)} \right\|_{U_j^{\gamma-2}} < \frac{1}{2^{\gamma-1}} .$$

Since $T_1^{-1}T_2 = M$ on U_{12}, B is trivial. Q.E.D.

CHAPTER 6

EXTENSION OF LOCALLY FREE SHEAVES ON HARTOGS' DOMAINS

In this chapter we prove the extension theorem for locally free sheaves on Hartogs' domains.

We need the following statement from the duality theory: Suppose $(X,\mathcal{O})$ is an n-dimensional complex manifold, Y is a relatively compact open subset of X, and $\mathcal{F}$ is a locally free sheaf on X. Suppose, for some $p \geq 0$, $H^p(X,\mathcal{F})$ and $H^p(Y,\mathcal{F})$ are Hausdorff and the map

$$H_c^{n-p}\left(Y, \mathcal{H}om_{\mathcal{O}}(\mathcal{F},\Omega)\right) \longrightarrow H_c^{n-p}\left(X, \mathcal{H}om_{\mathcal{O}}(\mathcal{F},\Omega)\right)$$

is surjective, where Ω is the sheaf of germs of holomorphic n-forms on X and $H_c{}^k$ denotes cohomology with compact supports. Then $H^{p+1}(X,\mathcal{F})$ is Hausdorff. (See the Appendix of Chapter 6 for a proof of this statement.)

In this chapter we use the following notations. If $0 < a$ in $\mathbb{R}^2$, then $Q(a)$ denotes $(\mathbb{P}_1 \times \mathbb{P}_1) - \overline{\Delta^2(a)}$ (where

$$\Delta^2(a) \subset \mathbb{C}^2 \subset \mathbb{P}_1 \times \mathbb{P}_1$$

in the natural way). For $\alpha \geq 0$ in $\mathbb{R}$, $L(\alpha)$ denotes the subset

$$\{z \in \mathbb{C} \,\big|\, |z| > \alpha\} \cup \{\infty\}$$

of $\mathbb{P}_1$.

(6.1) Lemma. Suppose $0 < a$ in $\mathbb{R}^2$ and $\mathcal{F}$ is a locally free sheaf on $Q(a)$. Then $H^1(Q(a),\mathcal{F})$ is Hausdorff.

Proof. Without loss of generality we can assume that $a_1 = a_2 = 1$. Let $\mathcal{G} = \mathcal{H}om_{\mathcal{O}}(\mathcal{F},\Omega)$, where $\mathcal{O}$ is the structure sheaf of $Q(a)$ and Ω is the sheaf of germs of holomorphic 2-forms on $Q(a)$. It suffices to show that

$$H_c{}^2(Q(2,2),\mathcal{G}) \longrightarrow H^c{}_2(Q(a),\mathcal{G})$$

is surjective. For this we need only show that the following two maps are surjective:

$$\alpha\colon H_c{}^2(Q(2,1),\mathcal{G}) \longrightarrow H_c{}^2(Q(a),\mathcal{G})$$

$$\beta\colon H_c{}^2(Q(2,2),\mathcal{G}) \longrightarrow H_c{}^2(Q(2,1),\mathcal{G}).$$

We prove only the surjectivity of α, because the proof for the surjectivity of β is completely analogous. We have the following exact sequence

$$H_c{}^2(Q(2,1),\mathcal{G}) \to H_c{}^2(Q(a),\mathcal{G}) \to H_c{}^2((\overline{\Delta(2)}-\overline{\Delta})\times\overline{\Delta},\mathcal{G}).$$

Since $\mathcal{G} \approx \mathcal{O}^r$ on $(\overline{\Delta(2)}-\overline{\Delta})\times\overline{\Delta}$ for some r (see the Appendix of Chapter 5), it suffices to show that

$$H_c{}^2((\overline{\Delta(2)}-\overline{\Delta})\times\overline{\Delta},\mathcal{O}) = 0 .$$

Take a C^∞ form $a d\bar{z}_1 \wedge d\bar{z}_2$ defined on $(\Delta(\lambda)-\overline{\Delta})\times\Delta(\mu)$ for some $\lambda > 2$ and $\mu > 1$ such that

$$\operatorname{Supp} a \subset (\Delta(\lambda)-\Delta(\nu)) \times \Delta(\mu)$$

for some $1 < \nu < \lambda$. Take $1 < \mu' < \mu$. Define a C^∞ function b on $(\Delta(\lambda)-\overline{\Delta})\times\Delta(\mu')$ as follows:

$$b(z_1,z_2) = \frac{1}{2\pi i}\int_{|\zeta|<\mu'} \frac{a(z_1,\zeta)d\zeta\wedge d\bar{\zeta}}{\zeta - z_2} .$$

Then $\dfrac{\partial b}{\partial \bar{z}_2} = a$ on $(\Delta(\lambda)-\bar{\Delta})\times\Delta(\mu')$ and

$$\operatorname{Supp} b \subset (\Delta(\lambda)-\Delta(\nu)) \times \Delta(\mu') .$$

It follows that

$$\bar{\partial}(-b d\bar{z}_1) = a d\bar{z}_1 \wedge d\bar{z}_2 .$$

Hence

$$H_c^2\left((\bar{\Delta}(2)-\bar{\Delta})\times\bar{\Delta}, \mathcal{O}\right) = 0 . \qquad \text{Q.E.D.}$$

(6.2) Lemma. *Suppose* $0 < \tilde{a} < a < c < b$ *in* $\mathbb{R}^2$ *and* $\mathcal{F}$ *is a locally free sheaf on* $Q(\tilde{a})$. *Let* $\mathfrak{U} = \{U_j\}_{j=1}^3$, *where* $U_1 = L(a_1)\times\Delta(b_2)$, $U_2 = \Delta(b_1)\times L(a_2)$, *and* $U_3 = L(c_1)\times L(c_2)$. *Then* $H^1_{L^2}(\mathfrak{U},\mathcal{F})$ *is Hausdorff.*

Proof. Since

$$\theta : H^1_{L^2}(\mathfrak{U},\mathcal{F}) \longrightarrow H^1(\mathfrak{U},\mathcal{F})$$

is continuous and $H^1(\mathfrak{U},\mathcal{F})$ is Hausdorff, it suffices to show that θ is injective. Suppose

$$\xi = \{\xi_{ij}\} \in Z^1_{L^2}(\mathfrak{U},\mathcal{F})$$

and $\xi = \delta\eta$ for some

$$\eta = \{\eta_i\} \in C^0(\mathfrak{U},\mathcal{F}) .$$

We claim that $\eta \in C^0_{L^2}(\mathfrak{U},\mathcal{F})$.

Take $b < \tilde{b}$ in $\mathbb{R}^2$. Let

$$\tilde{U}_1 = L(\tilde{a}_1) \times \Delta(\tilde{b}_2) ,$$

$$\tilde{U}_2 = \Delta(\tilde{b}_1) \times L(\tilde{a}_2) ,$$

and

$$\tilde{U}_3 = L(\tilde{a}_1) \times L(\tilde{a}_2) .$$

There exists a sheaf-isomorphism $\varphi_j : \mathcal{O}^r \longrightarrow \mathcal{F}$ on $\tilde{U}_j$ $(1 \leq j \leq 3)$, where $\mathcal{O}$ is the structure sheaf of $\mathbb{P}_1 \times \mathbb{P}_1$. Take a C^∞ volume element ω_j on $\tilde{U}_j$ $(1 \leq j \leq 3)$. We will show that $\varphi_j^{-1}(\eta_j)$ is an r-tuple of L^2 functions on U_j with respect to the volume element ω_j $(1 \leq j \leq 3)$. We prove only the case $j = 1$, because the cases $j = 2,3$ are completely analogous. Since $G^1(a_1,c_1) \times G^1(c_2,b_2)$ is relatively compact on $G^1(\tilde{a}_1,b_1) \times G^1(a_2,\tilde{b}_2)$ and $\varphi_1^{-1}(\eta_2)$ defines an r-tuple of holomorphic functions on $G^1(\tilde{a}_1,b_1) \times G^1(a_2,\tilde{b}_2)$, it follows that $\varphi_1^{-1}(\eta_2)$ defines an r-tuple of L^2 functions on $G^1(a_1,c_1) \times G^1(c_2,b_2)$. Since $\varphi_1^{-1}(\xi_{12})$ defines an r-tuple of L^2 functions on $G^1(a_1,c_1) \times G^1(c_2,b_2)$, it follows from $\eta_1 = \eta_2 - \xi_{12}$ that $\varphi_1^{-1}(\eta_1)$ defines an r-tuple of L^2 functions on $G^1(a_1,c_1) \times G^1(c_2,b_2)$. By applying (3.1) twice, we conclude that $\varphi_1^{-1}(\eta_1)$ is an r-tuple of L^2 functions on U_1 with respect to the volume element ω_1 .

Q.E.D.

For a Banach space E we denote by ${}_n\mathcal{O}(E)$ the sheaf of germs of E-valued holomorphic functions on $\mathbb{C}^n$.

Suppose D is an open subset of $\mathbb{C}^n$ and B is a subset of $D \times E$. We say that B (with the structure induced by $D \times E$) is a holomorphic vector-bundle of finite rank if for every $t \in D$ there exist an open neighborhood U of t in D and a holomorphic bundle-monomorphism $\varphi: U \times \mathbb{C}^r \longrightarrow U \times E$ such that $\operatorname{Im} \varphi = B \cap (U \times E)$. In such a case, if F is a complementary (closed) subspace of $\operatorname{Im} \varphi(t)$ in E, then for some open neighborhood W of t in U the map

$$\psi: (W \times \mathbb{C}^r) \oplus (W \times F) \longrightarrow W \times E$$

defined by φ and $W \times F \hookrightarrow W \times E$ is a holomorphic bundle-isomorphism, because $\psi(t): \mathbb{C}^r \oplus F \longrightarrow E$ is an isomorphism. So, if we give $B|W := B \cap (W \times E)$ the trivial bundle structure of $W \times \mathbb{C}^r$, then $W \times E \approx (B|W) \oplus (W \times F)$.

(6.3) Lemma. *Suppose E, F are Banach spaces, D is an open neighborhood of 0 in $\mathbb{C}^n$, and $\varphi: D \times E \longrightarrow D \times F$ is a holomorphic bundle-homomorphism. Suppose $\dim_{\mathbb{C}} \operatorname{Ker} \varphi(0) < \infty$ and $\operatorname{Im} \varphi(0)$ is complemented in F. Then there exists an open neighborhood U of 0 in D satisfying the following two conditions.*

(i) *The kernel of the sheaf-homomorphism ${}_n\mathcal{O}(E)|U \longrightarrow {}_n\mathcal{O}(F)|U$ induced by φ is coherent on U.*

(ii) *If W is the largest open subset of U such that $(\operatorname{Ker} \varphi)|W$ is a holomorphic vector bundle of finite rank, then $U - W$ is a subvariety of codimension ≥ 1 in U.*

Proof. (a) Consider first the very special case where $E = \mathbb{C}^p$ and $F = \mathbb{C}^q$. Set $U = D$. The first condition is clearly satisfied. To verify the second condition, we can assume that D is connected. Let $r = \max_{t \in D} \operatorname{rank} \varphi(t)$. Let W be the set of all $t \in D$ such that $\operatorname{rank} \varphi(t) = r$. Then W satisfies the requirement.

(b) Next consider the special case where $E = \mathbb{C}^p$. Let F^* be the set of all continuous linear functionals on F. For $f \in F^*$ define

$$\Psi_f : D \times E \longrightarrow D \times \mathbb{C}$$

by

$$\Psi_f(t,e) = \big(t, f(\varphi(t,e))\big) .$$

Ψ_f is represented by a p-tuple s_f of holomorphic functions on D. Let $\mathscr{A}$ be the subsheaf of ${}_n\mathcal{O}^p|D$ generated by $\{s_f\}_{f \in F^*}$. Let U be a relatively compact Stein open neighborhood of 0 in D. Since $\mathscr{A}$ is coherent, there exists a finite subset $\{f_1, \ldots, f_q\}$ of F^* such that $\{s_{f_i}\}_{i=1}^k$ generates $\mathscr{A}|U$. Since U is Stein, for every $f \in F^*$ there exist holomorphic functions $a_{f,i}$ on U $(1 \le i \le q)$ such that

$$(*) \qquad s_f = \sum_{i=1}^{q} a_{f,i} s_{f_i} \quad \text{on } U .$$

Let $\theta : U \times E \longrightarrow U \times \mathbb{C}^q$ be defined by the $q \times p$ matrix whose i^{th} row is s_{f_i} $(1 \le i \le q)$. It follows from (*) that $\operatorname{Ker} \theta = (\operatorname{Ker} \varphi)|U$. Hence the special case follows from (a).

(c) For the general case we have the decompositions $E = E_1 \oplus E_2$ and $F = F_1 \oplus F_2$, where $E_2 = \operatorname{Ker} \varphi(0)$ and

$F_1 = \operatorname{Im} \varphi(0)$. We can write

$$\varphi = \begin{pmatrix} \varphi_{11} & \varphi_{12} \\ \varphi_{21} & \varphi_{22} \end{pmatrix} ,$$

where

$$\varphi_{ij}: D \times E_j \longrightarrow D \times F_i$$

is a holomorphic bundle-homomorphism. Since $\varphi_{11}(0)$ is an isomorphism from E_1 to F_1 , there exists an open neighborhood $\tilde{U}$ of 0 in D such that, for $t \in \tilde{U}$, $\varphi_{11}(t)$ is an isomorphism from E_1 to F_1 .

For $t \in \tilde{U}$ and $e_j \in E_j$ $(j = 1,2)$,

$$(\dagger) \qquad \begin{cases} \varphi_{11}(t)e_1 + \varphi_{12}(t)e_2 = 0 \\ \varphi_{21}(t)e_1 + \varphi_{22}(t)e_2 = 0 \end{cases}$$

is equivalent to

$$(\#) \qquad \begin{cases} \left(\varphi_{22}(t) - \varphi_{21}(t)\varphi_{11}(t)^{-1}\varphi_{12}(t)\right) e_2 = 0 \\ e_1 = -\varphi_{11}(t)^{-1}\varphi_{12}(t)e_2 . \end{cases}$$

Define holomorphic bundle-homomorphisms

$$\psi: \tilde{U} \times E_2 \longrightarrow \tilde{U} \times F$$

$$\sigma: \tilde{U} \times E_2 \longrightarrow \tilde{U} \times E$$

by

$$\psi = \varphi_{22} - \varphi_{21}\varphi_{11}^{-1}\varphi_{12}$$

and

$$\sigma(t)e_2 = \left(-\varphi_{11}(t)^{-1}\varphi_{12}(t)e_2\right) \oplus e_2 .$$

Let $\tau: \tilde{U} \times E \longrightarrow \tilde{U} \times E_2$ be induced by the projection $E \longrightarrow E_2$ along E_1. By the equivalence of (†) and (#), we have $\sigma(\text{Ker } \psi) = (\text{Ker } \varphi)|\tilde{U}$ and $\tau((\text{Ker } \varphi)|\tilde{U}) = \text{Ker } \psi$. The general case follows from applying (b) to ψ. Q.E.D.

(6.4) Proposition. Suppose $0 < a$ in $\mathbb{R}^2$, D is an open polydisc in $\mathbb{C}^n$, and $\mathcal{F}$ is a locally free sheaf on $D \times Q(a)$. Then there exists a subvariety T of codimension ≥ 1 in D such that, for $t \in D-T$, the map
$\Gamma(D \times Q(a), \mathcal{F}) \longrightarrow \Gamma(\{t\} \times Q(a), \mathcal{F}(t))$ is surjective.

Proof. Take $a < a' < c' < b' < b$ in $\mathbb{R}^2$. Let $\tilde{U}_1 = L(a_1) \times \Delta(b_2)$, $\tilde{U}_2 = \Delta(b_1) \times L(a_2)$, and $U_3 = L(a_1) \times L(a_2)$. There exists a sheaf-isomorphism $\alpha_j: \mathcal{O}^r \longrightarrow \mathcal{F}$ on $D \times \tilde{U}_j$ $(1 \leq j \leq 3)$, where $\mathcal{O}$ is the structure sheaf of $\mathbb{C}^n \times \mathbb{P}_1 \times \mathbb{P}_1$. For $t \in D$ let $\alpha_j(t): \mathcal{O}^r(t) \longrightarrow \mathcal{F}(t)$ be induced by α_j $(1 \leq j \leq 3)$ and we identify $\mathcal{O}^r(t)$ with the sheaf of germs of holomorphic functions on $\mathbb{P}_1 \times \mathbb{P}_1$.

First we show that the restriction maps

$$\theta_t: \Gamma(\{t\} \times Q(a), \mathcal{F}(t)) \longrightarrow \Gamma(\{t\} \times Q(a'), \mathcal{F}(t)) \qquad (t \in D)$$
$$\theta: \Gamma(D \times Q(a), \mathcal{F}) \longrightarrow \Gamma(D \times Q(a'), \mathcal{F})$$

are bijective. Because of the sheaf-isomorphisms α_j, by (1.2) the restriction maps

$$\Gamma(\{t\} \times \tilde{U}_j, \mathcal{F}(t)) \longrightarrow \Gamma(\{t\} \times (Q(a') \cap \tilde{U}_j), \mathcal{F}(t))$$
$$\Gamma(D \times \tilde{U}_j, \mathcal{F}) \longrightarrow \Gamma(D \times (Q(a') \cap \tilde{U}_j), \mathcal{F})$$

are bijective $(j = 1,2)$. It follows that θ_t and θ are bijective.

Since $Q(a') \subset\subset Q(a)$, θ_t is compact. It follows that

$$(*) \qquad \dim_{\mathbb{C}} \Gamma(\{t\} \times Q(a'), \mathcal{F}(t)) < \infty \quad (t \in D) .$$

Let $U_1 = L(a_1') \times \Delta(b_2')$, $U_2 = \Delta(b_1') \times L(a_2')$, and $U_3 = L(c_1') \times L(c_2')$. Take a C^∞ volume element ω_j on $\tilde{U}_j$ $(1 \leq j \leq 3)$. Let $H_{i_0 \ldots i_p}$ be the Hilbert space of all r-tuples of holomorphic functions on $U_{i_0} \cap \ldots \cap U_{i_p}$ which are L^2 on $U_{i_0} \cap \ldots \cap U_{i_p}$ with respect to the volume element ω_{i_0} . Set $E = \bigoplus_{1 \leq i \leq 3} H_i$ and $F = \bigoplus_{1 \leq i < j \leq 3} H_{ij}$. Define a holomorphic bundle-homomorphism $\varphi: D \times E \longrightarrow D \times F$ as follows. For $t \in D$ and $e = \{e_i\} \in E$, define $\varphi(t,e) = (t,f)$, where $f = \{f_{ij}\}$ and

$$f_{ij} = \alpha_i(t)^{-1}(\alpha_j(t)e_j - \alpha_i(t)e_i) .$$

It follows from (*) and the bijectivity of θ_t that $\dim \operatorname{Ker} \varphi(t) < \infty$ for every $t \in D$. By (6.2), for every $t \in D$, $\operatorname{Im} \varphi(t)$ is closed in F and hence complemented in F . By (6.3) the kernel $\mathcal{K}$ of the sheaf-homomorphism ${}_n\mathcal{O}(E)|D \longrightarrow {}_n\mathcal{O}(F)|D$ induced by φ is coherent and there exists a subvariety T of codimension ≥ 1 in D such that $(\operatorname{Ker} \varphi)|D-T$ is a holomorphic vector bundle of finite rank. Since D is Stein, the restriction map

$$\Gamma(D,\mathcal{K}) \longrightarrow \operatorname{Ker} \varphi(t)$$

is surjective for every $t \in D-T$.

Because of the bijectivity of θ_t and θ, the natural maps

$$\Gamma(D,\mathcal{K}) \longrightarrow \Gamma(D \times Q(a), \mathcal{F})$$

and

$$\operatorname{Ker} \varphi(t) \longrightarrow \Gamma(\{t\} \times Q(a), \mathcal{F}(t))$$

are bijective $(t \in D)$. Hence the proposition follows. Q.E.D.

(6.5) Proposition. Suppose $0 < a$ in $\mathbb{R}^2$, D is an open polydisc in $\mathbb{C}^n$, A is a thick subset of D, and $\mathcal{F}$ is a locally free sheaf on $D \times Q(a)$ such that, for $t \in A$, $\mathcal{F}(t)$ can be extended to a coherent analytic sheaf on $\{t\} \times \mathbb{P}_1 \times \mathbb{P}_1$. Then $\mathcal{F}$ can be extended uniquely to a coherent analytic sheaf $\tilde{\mathcal{F}}$ on $D \times \mathbb{P}_1 \times \mathbb{P}_1$ such that $\tilde{\mathcal{F}}^{[n]} = \tilde{\mathcal{F}}$.

Proof. The map

$$([z_0, z_1], [w_0, w_1]) \longmapsto [z_0 w_0, z_0 w_1, z_1 w_0, z_1 w_1]$$

maps $\mathbb{P}_1 \times \mathbb{P}_1$ biholomorphically onto a complex submanifold M of $\mathbb{P}_3$. We identify $\mathbb{P}_1 \times \mathbb{P}_1$ with M through this biholomorphic map. Let B be the line bundle on $\mathbb{P}_3$ which is associated to the principal bundle $\mathbb{C}^4 - \{0\} \longrightarrow \mathbb{P}_3$. Let $\mathcal{L}$ be the sheaf of germs of holomorphic sections of $B^{-1} | \mathbb{P}_1 \times \mathbb{P}_1$, where B^{-1} is the dual bundle of B. Since every coherent analytic sheaf on $\mathbb{P}_1 \times \mathbb{P}_1$ can be trivially extended to a coherent analytic sheaf on $\mathbb{P}_3$, by the analog of Theorem B in algebraic geometry [7, pp.284-286], for every coherent analytic sheaf $\mathcal{G}$ on $\mathbb{P}_1 \times \mathbb{P}_1$ there exists $k_1 \geq 0$ such that

$$H^1(\mathbb{P}_1 \times \mathbb{P}_1, \mathcal{G} \otimes \mathcal{L}^k) = 0$$

for $k \geq k_1$. For $x \in \mathbb{P}_1 \times \mathbb{P}_1$ let $\mathfrak{m}_x$ be the ideal sheaf of the subvariety $\{x\}$ of $\mathbb{P}_1 \times \mathbb{P}_1$. The exact sequence

$$0 \longrightarrow \mathfrak{m}_x \mathcal{G} \otimes \mathcal{L}^k \longrightarrow \mathcal{G} \otimes \mathcal{L}^k \longrightarrow (\mathcal{G}/\mathfrak{m}_x \mathcal{G}) \otimes \mathcal{L}^k \longrightarrow 0$$

yields the exact sequence

$$\Gamma(\mathbb{P}_1 \times \mathbb{P}_1, \mathcal{G} \otimes \mathcal{L}^k) \longrightarrow \Gamma\big(\mathbb{P}_1 \times \mathbb{P}_1, (\mathcal{G}/\mathfrak{m}_x \mathcal{G}) \otimes \mathcal{L}^k\big)$$
$$\longrightarrow H^1(\mathbb{P}_1 \times \mathbb{P}_1, \mathfrak{m}_x \mathcal{G} \otimes \mathcal{L}^k) .$$

There exists $k_2 = k_2(\mathcal{G}, x)$ such that

$$H^1(\mathbb{P}_1 \times \mathbb{P}_1, \mathfrak{m}_x \mathcal{G} \otimes \mathcal{L}^k) = 0$$

for $k \geq k_2$. It follows from Nakayama's lemma that $\Gamma(\mathbb{P}_1 \times \mathbb{P}_1, \mathcal{G} \otimes \mathcal{L}^k)$ generates $(\mathcal{G} \otimes \mathcal{L}^k)_x$ for $k \geq k_2$. $\Gamma(\mathbb{P}_1 \times \mathbb{P}_1, \mathcal{L})$ generates $\mathcal{L}$, because the homogeneous linear functions on $\mathbb{C}^4$ induce a subset of $\Gamma(\mathbb{P}_1 \times \mathbb{P}_1, \mathcal{L})$ which generates $\mathcal{L}$. Since $\mathbb{P}_1 \times \mathbb{P}_1$ is compact, there exists $k_3 = k_3(\mathcal{G})$ such that $\Gamma(\mathbb{P}_1 \times \mathbb{P}_1, \mathcal{G} \otimes \mathcal{L}^k)$ generates $\mathcal{G} \otimes \mathcal{L}^k$ for $k \geq k_3$. Let $\tilde{\mathcal{L}}$ be the inverse image of $\mathcal{L}$ under the projection $D \times \mathbb{P}_1 \times \mathbb{P}_1 \longrightarrow \mathbb{P}_1 \times \mathbb{P}_1$. For $k \geq 0$ let $\mathcal{F}^{(k)} = \mathcal{F} \otimes \tilde{\mathcal{L}}^k$.

By (6.4) there exists a subvariety T_k of codimension ≥ 1 in D such that

$$\Gamma\big(D \times Q(a), \mathcal{F}^{(k)}\big) \longrightarrow \Gamma\big(\{t\} \times Q(a), \mathcal{F}^{(k)}(t)\big)$$

is surjective for $t \in D - T_k$. Since A is a thick subset of D, there exists

$$t^0 \in A - \bigcup_{k=0}^{\infty} T_k .$$

Since $\mathcal{F}(t^0)$ can be extended to a coherent analytic sheaf on $\mathbb{P}_1 \times \mathbb{P}_1$, there exists $k \geq 0$ such that $\Gamma(\{t^0\} \times Q(a), \mathcal{F}^{(k)}(t^0))$ generates $\mathcal{F}^{(k)}(t^0)$. By Nakayama's lemma, $\Gamma(D \times Q(a), \mathcal{F}^{(k)})$ generates $\mathcal{F}^{(k)}$ on $\{t^0\} \times Q(a)$. Take $a < b$ in $\mathbb{R}^2$. Then $\Gamma(D \times G^2(a,b), \mathcal{F})$ generates $\mathcal{F}$ on $\{t^0\} \times G^2(a,b)$. Let $\mathcal{R}$ be the subsheaf of $\mathcal{F}|D \times G^2(a,b)$ gnerated by $\Gamma(D \times G^2(a,b), \mathcal{F})$. $\mathcal{R}$ is coherent.

Take $a < a' < b' < b$ in $\mathbb{R}^2$ and take an arbitrary relatively compact connected open subset D' of D such that $A \cap D'$ is thick in D'. There exists a sheaf-epimorphism $\varphi: \mathcal{O}^p \longrightarrow \mathcal{R}$ on $D' \times G^2(a',b')$, where $\mathcal{O}$ is the structure sheaf of $\mathbb{C}^n \times \mathbb{P}_1 \times \mathbb{P}_1$. Since $0_{[n+1]\mathcal{R}} = 0$,

$$(\operatorname{Ker} \varphi)_{[n+1]\mathcal{O}^p} = \operatorname{Ker} \varphi .$$

By (4.6), $\operatorname{Ker} \varphi$ can be extended to a coherent analytic subsheaf $\mathcal{K}$ of $\mathcal{O}^p|D' \times \Delta^2(b')$ such that $\mathcal{K}_{[n+1]\mathcal{O}^p} = \mathcal{K}$. Let $\tilde{\mathcal{R}} = (\mathcal{O}^p/\mathcal{K})^{[n]}$. $\tilde{\mathcal{R}}$ is a coherent analytic sheaf on $D' \times \Delta^2(b')$ extending $\mathcal{R}_{[n]\mathcal{F}}|D' \times G^2(a',b')$.

Since $\tilde{\mathcal{R}}^{[n]} = \tilde{\mathcal{R}}$, by (3.17) there exists a thin set C in D' such that $\tilde{\mathcal{R}}(t)^{[0]} = \tilde{\mathcal{R}}(t)$ on $\{t\} \times \Delta^2(b')$ for $t \in D'-C$.

Let V be the subvariety of $D \times G^2(a,b)$ where $\mathcal{F}$ and $\mathcal{R}_{[n]\mathcal{F}}$ disagree. Every nonempty branch of V has dimension $\geq n+1$. Since $V \cap (\{t^0\} \times G^2(a,b)) = \emptyset$, there exists an open neighborhood U of t^0 in D such that $V \cap (U \times G^2(a',b')) = \emptyset$. By (2.19), $V|D \times G^2(a',b')$ can be extended to a subvariety $\tilde{V}$ of $D \times \Delta^2(b)$ whose every branch has dimension $\geq n+1$. By (2.17)(b), $\tilde{V}$ extends V.

Take a non-identically-zero holomorphic function f on $D \times \Delta^2(b)$ which vanishes identically on $\tilde{V}$. There exists $m \geq 0$ such that $f^m \mathcal{F} \subset \mathcal{R}_{[n]\mathcal{F}}$ on $D' \times G^2(a',b')$. By (3.14) and (3.15), for $t \in A \cap D' - C$,

$$\operatorname{Im}\left((f^m\mathcal{F})(t) \longrightarrow \tilde{\mathcal{R}}(t)\right) \mid \{t\} \times G^2(a',b')$$

can be extended to a coherent analytic subsheaf of $\tilde{\mathcal{R}}(t) \mid \{t\} \times \Delta^2(b')$. Since $(f^m\mathcal{F})_{[n]\tilde{\mathcal{R}}} = f^m\mathcal{F}$ on $D' \times G^2(a',b')$, by (4.4), $f^m\mathcal{F} \mid D' \times G^2(a',b')$ can be extended to a coherent analytic subsheaf of $\tilde{\mathcal{R}}$ on $D' \times \Delta^2(b')$. Since $\mathcal{F} \approx f^m\mathcal{F}$ on $D' \times G^2(a',b')$, $\mathcal{F} \mid D' \times G^2(a',b')$ can be extended to a coherent analytic sheaf on $D' \times \Delta^2(b')$. By (3.16) and the arbitrariness of D', $\mathcal{F}$ can be extended to a coherent analytic sheaf on $D \times \mathbb{P}_1 \times \mathbb{P}_1$. Q.E.D.

(6.6) Lemma. Suppose $0 < \alpha < \beta$ in $\mathbb{R}$ and r is a positive integer. Let $U_1 = \Delta(\beta)$, $U_2 = L(\alpha)$, and $U_{12} = U_1 \cap U_2$. Then there exists $\epsilon > 0$ satisfying the following: if D is an open subset of $\mathbb{C}^n$ and M is an $r \times r$ matrix of holomorphic functions on $D \times U_{12}$ with $\|I-M\|_{D \times U_{12}} < \epsilon$ (where I is the $r \times r$ identity matrix and $\|I-M\|_{D \times U_{12}}$ denotes the maximum of the sup norms on $D \times U_{12}$ of the entries of $I-M$), then there exists a nonsingular $r \times r$ matrix M_j of holomorphic functions on $D \times U_j$ $(j = 1,2)$ such that $M = M_1^{-1}M_2$ on $D \times U_{12}$.

Proof. For an open subset D of $\mathbb{C}^n$ and an $r \times r$ matrix A of holomorphic functions on $D \times U_{12}$ with Laurent series expansion

$$A = \sum_{\nu=-\infty}^{\infty} A_\nu z^\nu$$

(where z is the inhomogeneous coordinate of $\mathbb{P}_1$), then define an $r \times r$ matrix $e_j(A)$ of holomorphic functions on $D \times U_j$ $(j = 1,2)$ as follows:

$$e_1(A) = \sum_{\nu=0}^{\infty} A_\nu z^\nu$$

and

$$e_2(A) = \sum_{\nu=-\infty}^{-1} A_\nu z^\nu .$$

Let B_{12} (respectively B_1, B_2) be the Banach space of all uniformly bounded holomorphic functions on U_{12} (respectively U_1, U_2). We claim that, for $f \in B_{12}$, $e_j(f) \in B_j$ $(j = 1,2)$. Take $\alpha < \gamma < \beta$. Since $e_1(f) = f - e_2(f)$ on U_{12} and $G^1(\gamma,\beta) \subset\subset U_2$, $e_1(f)$ is uniformly bounded on $G^1(\gamma,\beta)$ and hence $e_1(f) \in B_1$. Likewise $e_2(f) \in B_2$. The claim is proved. By the closed graph theorem for Banach spaces, the map $B_{12} \longrightarrow B_j$ defined by $f \longrightarrow e_j(f)$ $(j = 1,2)$ is continuous. Hence there exists $C \geq 1$ such that

$$(*) \quad \|e_j(f)\|_{U_j} \leq C\|f\|_{U_{12}} \quad \text{for } f \in B_{12} \quad (j = 1,2) .$$

Take $0 < \epsilon < \dfrac{1}{4r^2C}$. We are going to show that ϵ satisfies the requirement. Observe that,

$$(\dagger)\quad \begin{cases} \text{if } A \text{ is an } r\times r \text{ scalar matrix with } \|I-A\| < C\epsilon , \\ \text{then } A^{-1} \text{ exists and } \|A^{-1}\| \le 2 , \end{cases}$$

because

$$A^{-1} = (I - (I-A))^{-1} = \sum_{\nu=0}^{\infty} (I-A)^{\nu} .$$

Take an open subset D of $\mathbb{C}^n$ and an $r \times r$ matrix M of holomorphic functions on $D \times U_{12}$ such that $\|I-M\|_{D\times U_{12}} < \epsilon$. Define, by induction on ν, $r\times r$ matrices $A^{(\nu)}$, $A_1^{(\nu)}$, $A_2^{(\nu)}$ of holomorphic functions respectively on $D\times U_{12}$, $D\times U_1$, $D\times U_2$ $(0 \le \nu < \infty)$ as follows:

(i) $A^{(0)} = M-I$

(ii) $A_j^{(\nu)} = (-1)^j e_j(A^{(\nu)})$

(iii) $A^{(\nu+1)} = A_1^{(\nu)} A^{(\nu)} (I+A_2^{(\nu)})^{-1}$.

We claim that

$$\|A^{(\nu)}\|_{D\times U_{12}} \le \frac{\epsilon}{2^{\nu}}$$

so that, by (*),

$$\|A_j^{(\nu)}\|_{D\times U_j} \le \frac{C\epsilon}{2^{\nu}}$$

$(j = 1,2)$ and, by $(\dagger)$, $(I+A_2^{(\nu)})^{-1}$ exists, making the definition of $A^{(\nu+1)}$ meaningful, and

$$\|(I+A_2^{(\nu)})^{-1}\|_{D\times U_2} \le 2 .$$

The claim follows from induction on ν and the following inequality coming from (iii):

$$\|A^{(\nu+1)}\|_{D\times U_{12}} \le r^2 \|A_1^{(\nu)}\|_{D\times U_1} \|A^{(\nu)}\|_{D\times U_{12}} \|(I+A_2^{(\nu)})^{-1}\|_{D\times U_2}$$

$$\le r^2 \cdot \frac{C\epsilon}{2^\nu} \cdot \frac{\epsilon}{2^\nu} \cdot 2 \le \frac{\epsilon}{2^{2\nu+1}} \le \frac{\epsilon}{2^{\nu+1}} .$$

It follows from (iii) and (ii) that

$$(I+A^{(\nu+1)})(I+A_2^{(\nu)}) = I + A_2^{(\nu)} + A^{(\nu+1)}(I+A_2^{(\nu)})$$

$$= I + A_2^{(\nu)} + A_1^{(\nu)}A^{(\nu)} = I + A_1^{(\nu)} + A^{(\nu)} + A_1^{(\nu)}A^{(\nu)}$$

$$= (I+A_1^{(\nu)})(I+A^{(\nu)}) .$$

Hence

$$I + A^{(\nu+1)} = (I+A_1^{(\nu)})(I+A^{(\nu)})(I+A_2^{(\nu)})^{-1} .$$

Let

$$M_j^{(\nu)} = (I+A_j^{(\nu)})(I+A_j^{(\nu-1)}) \dots (I+A_j^{(0)})$$

$(j = 1,2)$. Then

$$I + A^{(\nu+1)} = M_1^{(\nu)}(I+A^{(0)})(M_2^{(\nu)})^{-1} .$$

Let $M_j = \lim_{\nu \to \infty} M_j^{(\nu)}$ $(j = 1,2)$. Since

$$\|A_j^{(\nu)}\|_{D\times U_j} \le \frac{C\epsilon}{2^\nu}$$

for $0 \le \nu < \infty$, M_j is a nonsingular $r \times r$ matrix of holomorphic functions on $D \times U_j$ $(j = 1,2)$. Moreover, $M = M_1^{-1}M_2$ on $D \times U_{12}$. Q.E.D.

(6.7) Lemma. *Suppose D is an open subset of $\mathbb{C}^n$ and $\mathcal{F}$ is a locally free sheaf on $D \times \mathbb{P}_1$. Let $\pi: D \times \mathbb{P}_1 \to D$ be* *the natural projection. Then the zero*th *direct image* $\pi_*{}^0\mathcal{F}$ *of* $\mathcal{F}$ *under* π *is coherent on* D .

Proof. We can assume without loss of generality that D is an open polydisc. Take $0 < \tilde{a} < a < b < \tilde{b}$ in $\mathbb{R}$. Let $\tilde{U}_1 = \Delta(\tilde{b})$, $\tilde{U}_2 = L(\tilde{a})$, $U_1 = \Delta(b)$, $U_2 = L(a)$, and $\mathfrak{U} = \{U_1, U_2\}$. There exists a sheaf-isomorphism $\alpha_j: \mathcal{O}^r \to \mathcal{F}$ on $D \times \tilde{U}_j$ $(j = 1,2)$, where $\mathcal{O}$ is the structure sheaf of $\mathbb{C}^n \times \mathbb{P}_1$. For $t \in D$, let $\alpha_j(t): \mathcal{O}^r(t) \to \mathcal{F}(t)$ be induced by α_j $(j = 1,2)$ and we identify $\mathcal{O}^r(t)$ with the sheaf of germs of r-tuples of holomorphic functions on $\mathbb{P}_1$. Take a C^∞ volume element ω_j on $\tilde{U}_j$ $(j = 1,2)$.

First, we show that, for $t \in D$, the map

$$\theta_t: H^1_{L^2}(\mathfrak{U}, \mathcal{F}(t)) \longrightarrow H^1(\mathfrak{U}, \mathcal{F}(t))$$

is injective. Take $\xi \in Z^1_{L^2}(\mathfrak{U}, \mathcal{F}(t))$ and suppose $\xi = \delta\eta$ for some

$$\eta = \{\eta_1, \eta_2\} \in C^0(\mathfrak{U}, \mathcal{F}(t)) .$$

Take $a < c < b$ in $\mathbb{R}$. Since $G^1(c,b) \subset\subset G^1(a,\tilde{b})$ and $\alpha_1(t)^{-1}\eta_2$ defines an r-tuple of holomorphic functions on $G^1(a,\tilde{b})$, it follows that $\alpha_1(t)^{-1}\eta_2$ defines an r-tuple of L^2 holomorphic functions on $G^1(c,b)$. Since $\alpha_1(t)^{-1}\eta_1 = \alpha_1(t)^{-1}\eta_2 - \alpha_1(t)^{-1}\xi$ on $G^1(a,b)$, $\alpha_1(t)^{-1}\eta_1$ defines an r-tuple of L^2 holomorphic functions on $G^1(a,b)$. Hence $\alpha_1(t)^{-1}\eta_1$ is an r-tuple of L^2 holomorphic functions on U_1 . Likewise $\alpha_2(t)^{-1}\eta_2$ is an r-tuple of L^2 holomorphic functions on U_2 . Therefore $\eta \in C^0_{L^2}(\mathfrak{U}, \mathcal{F}(t))$ and θ_t is injective.

Since

$$\dim_{\mathbb{C}} H^1(\mathfrak{U},\mathcal{F}(t)) < \infty ,$$

it follows from the injectivity of θ_t that

$$\dim_{\mathbb{C}} H^1_{L^2}(\mathfrak{U},\mathcal{F}(t)) < \infty$$

and hence $H^1_{L^2}(\mathfrak{U},\mathcal{F}(t))$ is Hausdorff for $t \in D$.

Let $H_{i_0 \dots i_p}$ be the Hilbert space of all r-tuples of holomorphic functions on $U_{i_0} \cap \dots \cap U_{i_p}$ which are L^2 on $U_{i_0} \cap \dots \cap U_{i_p}$ with respect to the volume element ω_{i_0}. Set $E = H_1 \oplus H_2$ and $F = H_{12}$. Define a holomorphic bundle-homomorphism $\varphi : D \times E \longrightarrow D \times F$ as follows. For $t \in D$ and $e = \{e_1, e_2\} \in E$, define $\varphi(t,e) = (t,f)$, where

$$f = \alpha_1(t)^{-1}\alpha_2(t)e_2 - e_1 .$$

Since $\dim_{\mathbb{C}} \Gamma(\mathfrak{U},\mathcal{F}(t)) < \infty$, $\dim_{\mathbb{C}} \operatorname{Ker} \varphi(t) < \infty$ for $t \in D$. Since $H^1_{L^2}(\mathfrak{U},\mathcal{F}(t))$ is Hausdorff, $\operatorname{Im} \varphi(t)$ is closed in F and hence complemented in F $(t \in D)$. By (6.3) the kernel $\mathcal{K}$ of the sheaf-homomorphism ${}_n\mathcal{O}(E)|D \longrightarrow {}_n\mathcal{O}(F)|D$ induced by φ is coherent. Since $\pi_*^0\mathcal{F}$ is isomorphic to $\mathcal{K}$, the lemma follows. Q.E.D.

(6.8) <u>Proposition</u>. <u>Suppose</u> D <u>is a domain in</u> $\mathbb{C}^n$ <u>and</u> $\mathcal{F}$ <u>is a locally free sheaf of rank</u> r <u>on</u> $D \times \mathbb{P}_1$. <u>Let</u> A <u>be the subset of</u> D <u>consisting of all</u> t <u>such that</u> $\mathcal{F}(t) \approx \mathcal{O}^r$ <u>on</u> $\{t\} \times \mathbb{P}_1$, <u>where</u> $\mathcal{O}$ <u>is the structure sheaf of</u> $\mathbb{P}_1$. <u>Then</u>

there exists a subvariety T *in* D *such that*

(i) $A \cap T = \emptyset$,

(ii) *for every open subset* W *of* $D-T$ *which is biholomorphic to either a polydisc or the product of an annulus and a polydisc,* $\mathcal{F} \approx \hat{\mathcal{O}}^r$ *on* $W \times \mathbb{P}_1$, *where* $\hat{\mathcal{O}}$ *is the structure sheaf of* $\mathbb{C}^n \times \mathbb{P}_1$.

Proof. When $A = \emptyset$, we can set $T = D$. So we assume that $A \neq \emptyset$.

Let $\mathcal{R}$ be the zeroth direct image of $\mathcal{F}$ under the natural projection $\pi: D \times \mathbb{P}_1 \to D$. By (6.7), $\mathcal{R}$ is coherent on D . Let T_1 be the subvariety of D where $\mathcal{R}$ is not locally free.

Let $\mathcal{G}$ be the subsheaf of $\mathcal{F}$ generated by $\mathcal{R}$, i.e. $s \in \mathcal{G}_x$ if and only if there exist $v_1, \ldots, v_p \in \Gamma(U \times \mathbb{P}_1, \mathcal{F})$ for some open neighborhood U of $\pi(x)$ such that $s \in (\sum_{i=1}^{p} \hat{\mathcal{O}} v_i)_x$. $\mathcal{G}$ is coherent. Let T_2 be the subvariety of $D \times \mathbb{P}_1$ where $\mathcal{G}$ disagrees with $\mathcal{F}$.

Let $T = T_1 \cup \pi(T_2)$. We claim that T satisfies the requirement. Take $t^0 \in A$. Take $0 < \tilde{a} < a < b < \tilde{b}$ in $\mathbb{R}$. Let $\tilde{U}_1 = \Delta(\tilde{b})$, $\tilde{U}_2 = L(\tilde{a})$, $U_1 = \Delta(b)$, and $U_2 = L(a)$. Set $\tilde{U}_{12} = \tilde{U}_1 \cap \tilde{U}_2$ and $U_{12} = U_1 \cap U_2$. Let G be an open polydisc neighborhood of t^0 in D . There exists a sheaf-isomorphism $\alpha_j: \hat{\mathcal{O}}^r \to \mathcal{F}$ on $G \times \tilde{U}_j$ $(j = 1,2)$. Let M be the nonsingular $r \times r$ matrix of holomorphic functions on $G \times \tilde{U}_{12}$ which represents $\alpha_1^{-1}\alpha_2: \hat{\mathcal{O}}^r \to \hat{\mathcal{O}}^r$. Since $\mathcal{F}(t^0) \approx \mathcal{O}^r$ on $\{t^0\} \times \mathbb{P}_1$, there exists a nonsingular $r \times r$ matrix M_j

of holomorphic functions on $\{t^0\} \times \tilde{U}_j$ $(j = 1,2)$ such that $M = M_1^{-1}M_2$ on $\{t^0\} \times \tilde{U}_{12}$. By replacing M by $(M_1 \circ \sigma)M(M_2 \circ \sigma)^{-1}$ (where

$$\sigma: G \times \tilde{U}_{12} \longrightarrow \{t^0\} \times \tilde{U}_{12}$$

is induced by the map $G \longrightarrow \{t^0\}$), we can assume that $M = I$ on $\{t^0\} \times \tilde{U}_{12}$, where I is the identity matrix..

Since for every $\epsilon > 0$ there exists an open neighborhood H_ϵ of t^0 in G such that $\|I-M\|_{H_\epsilon \times U_{12}} < \epsilon$, by (6.6) for some open neighborhood H of t^0 in G there exists a nonsingular $r \times r$ matrix P_j of holomorphic functions on $H \times U_j$ $(j = 1,2)$ such that $M = P_1^{-1}P_2$ on $H \times U_{12}$. Hence $\mathcal{F} \approx \hat{\mathcal{O}}^r$ on $H \times \mathbb{P}_1$. It follows that $H \cap T = \emptyset$ and $\mathcal{R} \approx {}_n\mathcal{O}^r$ on H.

Suppose W is a Stein open subset of $D-T$ which is biholomorphic to either a polydisc or the product of an annulus and a polydisc. Since $\mathcal{R}$ is locally free of rank r on $D-T_1$, $\mathcal{R} \approx {}_n\mathcal{O}^r$ on W. $\mathcal{R}$ is generated on $W \times \mathbb{P}_1$ by r elements of $\Gamma(W \times \mathbb{P}_1, \mathcal{F})$. Hence $\mathcal{F} \approx \hat{\mathcal{O}}^r$ on $W \times \mathbb{P}_1$. Q.E.D.

Suppose G_1, G_2 are open subsets of a complex manifold $(X,\mathcal{O})$ and M is a nonsingular $r \times r$ matrix of holomorphic functions on an open subset of X containing $G_1 \cap G_2$. Then $\mathcal{L}_M(G_1,G_2)$ denotes the locally free sheaf of rank r on $G_1 \cap G_2$ characterized as follows. There exists a sheaf-isomorphism $\varphi_j: \mathcal{O}^r \longrightarrow \mathcal{L}_M(G_1,G_2)$ on G_j $(j = 1,2)$ such that the sheaf-isomorphism $\varphi_1^{-1}\varphi_2: \mathcal{O}^r \longrightarrow \mathcal{O}^r$ on $G_1 \cap G_2$ is represented by M (when elements of $\mathcal{O}_x{}^r$ are written as column vectors).

(6.9) Proposition. Suppose $0 < a < b$ in $\mathbb{R}^2$, D is an open polydisc in $\mathbb{C}^n$, and $\mathcal{F}$ is a locally free sheaf of rank r on $D \times G^2(a,b)$. Suppose $t^0 \in D$ and $\mathcal{F}(t^0)$ can be extended to a coherent analytic sheaf on $\{t^0\} \times \Delta^2(b)$. Then there exists a closed thin set T in D satisfying the following: There exist

(i) an open polydisc neighborhood U of t^0 in $D-T$,

(ii) $a_1 < a_1' < b_1' < b_1$ in $\mathbb{R}$,

(iii) $a_2' > a_2$ in $\mathbb{R}$,

(iv) a locally free sheaf $\mathcal{A}$ of rank r on $U \times (\Delta(b_1') \times \mathbb{P}_1 - \overline{\Delta^2(a_1', a_2)})$,

(v) a locally free sheaf $\mathcal{J}$ of rank r on $U \times Q(a_1', a_2')$,

such that

(a) $\mathcal{F}$ is isomorphic to $\mathcal{A}$ on $U \times G^2((a_1', a_2), (b_1', b_2))$,

(b) $\mathcal{A}$ is isomorphic to $\mathcal{J}$ on $U \times (\Delta(b_1') \times \mathbb{P}_1 - \overline{\Delta^2(a_1', a_2')})$.

Proof. There exists a nonsingular $r \times r$ matrix S of holomorphic functions on $D \times G^1(a_1,b_1) \times G^1(a_2,b_2)$ such that

$$\mathcal{F} \approx \mathcal{L}_S\big(D \times G^1(a_1,b_1) \times \Delta(b_2),\ D \times \Delta(b_1) \times G^1(a_2,b_2)\big) .$$

$\mathcal{F}(t^0)$ can be extended to a coherent analytic sheaf $\mathcal{F}(t^0)^\sim$ on $\{t^0\} \times \Delta^2(b)$. We can assume that $\big(\mathcal{F}(t^0)^\sim\big)^{[0]} = \mathcal{F}(t^0)^\sim$. By (3.13), $\mathcal{F}(t^0)^\sim$ is locally free and hence is isomorphic to ${}_2\mathcal{O}^r$ on $\{t^0\} \times \Delta^2(b)$. We can therefore choose S so that $S = I$ on $\{t^0\} \times G^1(a_1,b_1) \times G^1(a_2,b_2)$, where I is the identity matrix.

Consider the following locally free sheaf of rank r on $D \times G^1(a_1,b_1) \times \mathbb{P}_1$:

$$\mathcal{R} := \mathcal{L}_S\left(D \times G^1(a_1,b_1) \times \Delta(b_2),\ D \times G^1(a_1,b_1) \times L(a_2)\right) .$$

Since $S = I$ on $\{t^0\} \times G^1(a_1,b_1) \times G^1(a_2,b_2)$, by (6.8) there exists a subvariety Z of codimension ≥ 1 in $D \times G^1(a_1,b_1)$ such that, for every open subset W of $D \times G^1(a_1,b_1)-Z$ which is biholomorphic to the product of an annulus and a polydisc, $\mathcal{R}$ is isomorphic to $\tilde{\mathcal{O}}^r$ on $W \times \mathbb{P}_1$, where $\tilde{\mathcal{O}}$ is the structure sheaf of $\mathbb{C}^n \times \mathbb{P}_1 \times \mathbb{P}_1$.

Let $\pi: D \times G^1(a_1,b_1) \longrightarrow D$ be the natural projection. Let A be the subset of Z where the rank of $\pi|Z$ is $\leq n-1$. Take $a_1 < a_1^* < b_1^* < b_1$ in $\mathbb{R}$. Let $T = \pi\left(A \cap \left(D \times \overline{G^1(a_1^*,b_1^*)}\right)\right)$. We claim that T satisfies the requirement.

Take $t \in D-T$. $\left(\{t\} \times \overline{G^1(a_1^*,b_1^*)}\right) \cap Z$ is at most discrete. We can find $a_1^* < a_1' < b_1' < b_1^*$ in $\mathbb{R}$ such that $\{t\} \times \overline{G^1(a_1',b_1')}$ is disjoint from Z . There exists an open polydisc neighborhood $\hat{U}$ of t in $D-T$ such that $\hat{U} \times G^1(a_1',b_1')$ is disjoint from Z . $\mathcal{R}$ is therefore isomorphic to $\tilde{\mathcal{O}}^r$ on $\hat{U} \times G^1(a_1',b_1') \times \mathbb{P}_1$. Hence there exist nonsingular $r \times r$ matrices P and P' of holomorphic functions on $\hat{U} \times G^1(a_1',b_1') \times L(a_2)$ and $\hat{U} \times G^1(a_1',b_1') \times \Delta(b_2)$ respectively such that $S = (P')^{-1}P$ on $\hat{U} \times G^1(a_1',b_1') \times G^1(a_2,b_2)$. Let $\hat{P} = P \circ \sigma$, where

$$\sigma:\ \hat{U} \times G^1(a_1',b_1') \times \mathbb{P}_1 \longrightarrow \hat{U} \times G^1(a_1',b_1') \times \{\infty\}$$

is induced by $\mathbb{P}_1 \longrightarrow \{\infty\}$. By replacing P and P'

respectively by $\hat{P}^{-1}P$ and $\hat{P}^{-1}P'$, we can assume without loss of generality that $P = I$ on $\hat{U} \times G^1(a_1', b_1') \times \{\infty\}$.

Consider the following locally free sheaf of rank r on $\hat{U} \times \mathbb{P}_1 \times L(a_2)$:

$$\mathcal{G} := \mathcal{L}_P\big(\hat{U} \times L(a_1') \times L(a_2),\ \hat{U} \times \Delta(b_1') \times L(a_2)\big) .$$

Let

$$p: \hat{U} \times \mathbb{P}_1 \times L(a_2) \longrightarrow \hat{U} \times L(a_2)$$

be the natural projection. For $s \in \hat{U} \times \{\infty\}$, $\mathcal{G}(s) \approx \mathcal{O}^r$ on $p^{-1}(s)$, where $\mathcal{O}$ is the structure sheaf of $\mathbb{P}_1$. By (6.8), there exists an open neighborhood B of $\hat{U} \times \{\infty\}$ in $\hat{U} \times L(a_2)$ such that, for any open polydisc B' contained in B, $\mathcal{G}$ is isomorphic to $\tilde{\mathcal{O}}^r$ on $p^{-1}(B')$.

Choose $a_2' > a_2$ in $\mathbb{R}$ and an open polydisc neighborhood U of t in $\hat{U}$ such that $U \times L(a_2') \subset B$. $\mathcal{G}$ is isomorphic to $\tilde{\mathcal{O}}^r$ on $U \times \mathbb{P}_1 \times L(a_2')$. It follows that there exist nonsingular $r \times r$ matrices M and M' of holomorphic functions on $U \times L(a_1') \times L(a_2')$ and $U \times \Delta(b_1') \times L(a_2')$ respectively such that $P = M(M')^{-1}$ on $U \times G^1(a_1', b_1') \times L(a_2')$.

Define

$$\mathcal{A} = \mathcal{L}_P\big(U \times G^1(a_1', b_1') \times \mathbb{P}_1,\ U \times \Delta(b_1') \times L(a_2)\big)$$

and

$$\mathcal{J} = \mathcal{L}_M\big(U \times L(a_1') \times \mathbb{P}_1,\ U \times \mathbb{P}_1 \times L(a_2')\big) .$$

Since $P = P'S$ on $U \times G^1(a_1', b_1') \times G^1(a_2, b_2)$, $\mathcal{A}$ is isomorphic to $\mathcal{J}$ on $U \times G^2\big((a_1', a_2), (b_1', b_2)\big)$. Since $M = PM'$ on

$U \times G^1(a_1',b_1') \times L(a_2')$, $\mathcal{J}$ is isomorphic to $\mathcal{A}$ on $U \times \left(\Delta(b_1') \times \mathbb{P}_1 - \overline{\Delta^2(a_1',a_2')}\right)$. Q.E.D.

(6.10) Theorem. Suppose $0 \leq a < b$ in $\mathbb{R}^2$, D is a domain in $\mathbb{C}^n$, and $\mathcal{F}$ is a locally free sheaf of rank r on $D \times G^2(a,b)$. Suppose A is a thick set in D and, for every $t \in A$, $\mathcal{F}(t)$ can be extended to a coherent analytic sheaf on $\{t\} \times \Delta^2(b)$. Then $\mathcal{F}$ can be extended uniquely to a coherent analytic sheaf $\tilde{\mathcal{F}}$ on $D \times \Delta^2(b)$ satisfying $\tilde{\mathcal{F}}^{[n]} = \tilde{\mathcal{F}}$.

Proof. (a) Consider first the special case where D is an open polydisc. There exists a closed thin set T in D satisfying the conditions of (6.9). Let D' be the largest open subset of D-T such that $\mathcal{F}|D' \times G^2(a,b)$ can be extended to a coherent analytic sheaf on $D' \times \Delta^2(b)$. Let B be the set of all $t \in D-T$ such that $\mathcal{F}(t)$ can be extended to a coherent analytic sheaf on $\{t\} \times \Delta^2(b)$. Let C be the set of all $t \in D-T$ such that, for every open neighborhood W of t in D-T , $W \cap B$ is thick in W . Since $A-T \subset B$ and A-T is thick in D-T , C is thick in D-T .

We claim that $D' = (D-T) \cap \overline{C}$. Clearly $D' \subset (D-T) \cap \overline{C}$. Take $t \in (D-T) \cap \overline{C}$. There exist U and $\mathcal{J}$ satisfying the conditions of (6.9). Since $U \cap C$ is thick in U and, for $t \in C$, $\mathcal{J}(t)$ can be extended to a coherent analytic sheaf on $\{t\} \times \mathbb{P}_1 \times \mathbb{P}_1$, by (6.5) $\mathcal{J}$ can be extended to a coherent analytic sheaf on $U \times \mathbb{P}_1 \times \mathbb{P}_1$. It follows that $\mathcal{F}|U \times G^2(a,b)$ can be extended to a coherent analytic sheaf on $U \times \Delta^2(b)$. Hence $U \subset D'$ and the claim is proved.

Since D' is nonempty and is both open and closed in the connected set $D-T$, it follows that $D' = D-T$. Let $\mathcal{G}$ be the coherent analytic sheaf on $(D-T) \times \Delta^2(b)$ which extends $\mathcal{F}|(D-T) \times G^2(a,b)$ and satisfies $\mathcal{G}^{[n]} = \mathcal{G}$.

Take $t \in T$. It suffices to show that, for some open neighborhood W of t in D, $\mathcal{F}|W \times G^2(a,b)$ can be extended to a coherent analytic sheaf on $W \times \Delta^2(b)$. We can assume without loss of generality that $t = 0$ and there exist $0 < \gamma$ in $\mathbb{R}^{n-1}$ and $0 < \alpha < \beta$ in $\mathbb{R}$ such that $\Delta^{n-1}(\gamma) \times \Delta(\beta) \subset D$ and $(\Delta^{n-1}(\gamma) \times G^1(\alpha,\beta)) \cap T = \emptyset$. Let $\mathcal{R}$ be the sheaf on $(\Delta^{n-1}(\gamma) \times \Delta(\beta) \times G^2(a,b)) \cup (\Delta^{n-1}(\gamma) \times G^1(\alpha,\beta) \times \Delta^2(b))$ which agrees with $\mathcal{F}$ on $\Delta^{n-1}(\gamma) \times \Delta(\beta) \times G^2(a,b)$ and agrees with $\mathcal{G}$ on $\Delta^{n-1}(\gamma) \times G^1(\alpha,\beta) \times \Delta^2(b)$. $\mathcal{R}$ is coherent and $\mathcal{R}^{[n]} = \mathcal{R}$. By (5.8), $\mathcal{R}$ can be extended to a coherent analytic sheaf $\tilde{\mathcal{R}}$ on $\Delta^{n-1}(\gamma) \times \Delta(\beta) \times \Delta^2(b)$. $\tilde{\mathcal{R}}$ extends $\mathcal{F}|\Delta^{n-1}(\gamma) \times \Delta(\beta) \times G^2(a,b)$.

(b) For the general case let D' be the largest open subset of D such that $\mathcal{F}|D' \times G^2(a,b)$ can be extended to a coherent analytic sheaf on $D' \times \Delta^2(b)$. Take an open polydisc P in D such that $P \cap A$ is thick in P. By (a), $P \subset D'$. Hence $D' \neq \emptyset$. Because of (a), D' is closed in D. It follows that $D' = D$. Q.E.D.

(6.11) Corollary. *Suppose* $0 \leq a < b$ *in* $\mathbb{R}^2$, D *is a domain in* $\mathbb{C}^n$, *and* $\mathcal{F}$ *is a coherent analytic sheaf on* $D \times G^2(a,b)$ *such that* $\mathcal{F}^{[n]} = \mathcal{F}$. *Suppose* A *is a thick set in* D *and, for* $t \in A$, $\mathcal{F}(t)$ *can be extended to a coherent*

<u>analytic sheaf on</u> $\{t\} \times \Delta^2(b)$. <u>Then</u> $\mathcal{F}$ <u>can be extended uniquely to a coherent analytic sheaf</u> $\tilde{\mathcal{F}}$ <u>on</u> $D \times \Delta^2(b)$ <u>satisfying</u> $\tilde{\mathcal{F}}^{[n]} = \tilde{\mathcal{F}}$.

<u>Proof</u>. Let $V = S_{n+1}(\mathcal{F})$. By (3.13) $\dim V \leq n-1$. Take $a < a' < b' < b$ in $\mathbb{R}^2$. Let

$$\pi: D \times G^2(a,b) \longrightarrow D$$

be the natural projection and let

$$C = \pi\left(V \cap \left(D \times \overline{G^2(a',b')}\right)\right) .$$

C is a closed thin set in D . Let $D' = D - C$. By (6.10), $\mathcal{F}|D' \times G^2(a',b')$ can be extended to a coherent analytic sheaf on $D' \times \Delta^2(b')$.

(a) Consider first the special case where $\dim V \cap \pi^{-1}(t) \leq 0$ for $t \in D$. Let D^* be the largest open subset of D such that $\mathcal{F}|D^* \times G^2(a,b)$ can be extended to a coherent analytic sheaf on $D^* \times \Delta^2(b)$. Take a boundary point t of D^* in D . There exist $a' < a^* < b^* < b'$ in $\mathbb{R}^2$ and a connected open neighborhood U of t in D such that $V \cap (U \times G^2(a^*,b^*)) = \emptyset$. By (6.10), $\mathcal{F}|U \times G^2(a^*,b^*)$ can be extended to a coherent analytic sheaf on $U \times \Delta^2(b^*)$. Hence $U \subset D^*$. Since D^* is both open and closed in D , $D^* = D$.

(b) Let L be an arbitrary relatively compact open subset of D . By (2.2) there exists a 2-dimensional plane T in $\mathbb{C}^n \times \mathbb{C}^2$ such that for some nonempty open subset Q of L and for some connected open neighborhood R of L in D we have

(i) $(Q+T) \cap (\mathbb{C}^n \times \Delta^2(b)) \subset D' \times \Delta^2(b)$,

(ii) $L \times \Delta^2(b) \subset (R+T) \cap (\mathbb{C}^n \times \Delta^2(b)) \subset D \times \Delta^2(b)$,

(iii) $\dim(x+T) \cap V \leq 0$ for $x \in V$,

where

$$Q+T = \{x+y \mid x \in Q,\ y \in T\}$$

and

$$x+T = \{x+y \mid y \in T\} .$$

By (a), $\mathcal{F} \mid (R+T) \cap (\mathbb{C}^n \times G^2(a,b))$ can be extended to a coherent analytic sheaf $\mathcal{F}^*$ on $(R+T) \cap (\mathbb{C}^n \times \Delta^2(b))$. It follows that $\mathcal{F}^* \mid L \times \Delta^2(b)$ extends $\mathcal{F} \mid L \times G^2(a,b)$. The corollary follows from the arbitrariness of L . Q.E.D.

Appendix of Chapter 6

Duality

(6.A.1) Lemma. Suppose $(X,\mathcal{O})$ is an n-dimensional complex manifold, Ω is the sheaf of germs of holomorphic n-forms on X, $\mathfrak{U}$ is a countable Stein covering of X, and $\mathcal{F}$ is a locally free sheaf on X. Let D_p be the dual of the Frechet space $C^p(\mathfrak{U},\mathcal{F})$. Then $H_p(D.) \approx H_c^{n-p}(X, \mathcal{H}om_{\mathcal{O}}(\mathcal{F},\Omega))$, where $H_c^{\cdot}$ denotes cohomology groups with compact support.

Proof. Let $\mathcal{E}^{n,q}$ be the sheaf of germs of C^∞ (n,q)-forms on X. Let $\mathfrak{U} = \{U_i\}_{i=1}^{\infty}$. Denote by $C_p(\mathfrak{U}, hom_{\mathcal{O},c}(\mathcal{F},\mathcal{E}^{n,q}))$ the set of all

$$\{\xi_{i_0 \ldots i_p}\}_{1 \leq i_0 < \ldots < i_p}$$

such that

$$\xi_{i_0 \ldots i_p} \in \Gamma_c\left(U_{i_0} \cap \ldots \cap U_{i_p}, \mathcal{H}om_{\mathcal{O}}(\mathcal{F},\mathcal{E}^{n,q})\right)$$

and $\xi_{i_0 \ldots i_p} = 0$ except for a finite number of $(i_0,\ldots,i_p)$. Consider the following commutative diagram:

$$\begin{array}{ccccccccc}
 & & 0 & & 0 & & & & 0 \\
 & & \uparrow & & \uparrow & & & & \uparrow \\
0 & \longrightarrow & \Gamma_c(X, \mathcal{H}om_{\mathcal{O}}(\mathcal{F}, \mathcal{E}^{n,0})) & \longrightarrow & \Gamma_c(X, \mathcal{H}om_{\mathcal{O}}(\mathcal{F}, \mathcal{E}^{n,1})) & \longrightarrow \cdots \longrightarrow & \Gamma_c(X, \mathcal{H}om_{\mathcal{O}}(\mathcal{F}, \mathcal{E}^{n,n})) & \longrightarrow & 0 \\
 & & d\uparrow & & d\uparrow & & d\uparrow & & \uparrow \\
0 & \longrightarrow & C_0(\mathfrak{U}, hom_{\mathcal{O},c}(\mathcal{F}, \mathcal{E}^{n,0})) & \longrightarrow & C_0(\mathfrak{U}, hom_{\mathcal{O},c}(\mathcal{F}, \mathcal{E}^{n,1})) & \longrightarrow \cdots \longrightarrow & C_0(\mathfrak{U}, hom_{\mathcal{O},c}(\mathcal{F}, \mathcal{E}^{n,n})) & \xrightarrow{\alpha_0} & D_0 \to 0 \\
 & & d\uparrow & & d\uparrow & & d\uparrow & & \uparrow \\
0 & \longrightarrow & C_1(\mathfrak{U}, hom_{\mathcal{O},c}(\mathcal{F}, \mathcal{E}^{n,0})) & \longrightarrow & C_1(\mathfrak{U}, hom_{\mathcal{O},c}(\mathcal{F}, \mathcal{E}^{n,1})) & \longrightarrow \cdots \longrightarrow & C_1(\mathfrak{U}, hom_{\mathcal{O},c}(\mathcal{F}, \mathcal{E}^{n,n})) & \xrightarrow{\alpha_1} & D_1 \to 0 \\
 & & \uparrow & & \uparrow & & \uparrow & & \uparrow \\
 & & \vdots & & \vdots & & \vdots & & \vdots
\end{array}$$

where α_p is defined via the duality of $H_c{}^n(U_{i_0} \cap \ldots \cap U_{i_p}, \mathcal{H}om_{\mathcal{O}}(\mathcal{F}, \Omega))$ and $\Gamma(U_{i_0} \cap \ldots \cap U_{i_p}, \mathcal{F})$ and d is the boundary map. Since

$$H_c{}^q(U_{i_0} \cap \ldots \cap U_{i_p}, \mathcal{H}om_{\mathcal{O}}(\mathcal{F}, \Omega)) = 0$$

for $q < n$, all the rows except the first one are exact. Take a partition of unity $\{\rho_i\}$ subordinate to $\mathfrak{U}$. If

$$\xi = \{\xi_{i_0 \ldots i_p}\} \in C_p(\mathfrak{U}, hom_{\mathcal{O},c}(\mathcal{F}, \mathcal{E}^{n,q}))$$

and $d\xi = 0$, then define

$$\eta = \{\eta_{i_0 \ldots i_{p+1}}\} \in C_{p+1}(\mathfrak{U}, hom_{\mathcal{O},c}(\mathcal{F}, \mathcal{E}^{n,q}))$$

by

$$\eta_{i_0 \ldots i_{p+1}} = \sum_{\nu=0}^{p+1} (-1)^\nu \rho_{i_\nu} \xi_{i_0 \ldots i_{\nu-1} i_{\nu+1} \ldots i_{p+1}}$$

and it is straightforward to verify that $d\eta = \xi$. Hence all the columns except the last one are all exact. By chasing the diagram in the standard zigzag manner, we conclude that

$$H_p(D.) \approx H_c^{n-p}(X, \mathcal{H}om_{\mathcal{O}}(\mathcal{F},\Omega)) . \qquad \text{Q.E.D.}$$

(6.A.2) Lemma. *Suppose* E, F *are Fréchet spaces and* $S,T:E \longrightarrow F$ *are continuous linear maps. Suppose* S *is compact and* T *is injective and has closed image. Then* $S+T$ *has closed image.*

Proof. Since S is compact, there exists an open neighborhood U of 0 in E such that $S(U)$ is relatively compact in F.

We show first that $\text{Ker}(S+T)$ is finite-dimensional. Take $\{x_\nu\} \subset U \cap \text{Ker}(S+T)$. Since $S(U)^-$ is compact, there exists a subsequence $\{y_\nu\}$ of $\{x_\nu\}$ such that Sy_ν converges to some $y \in F$. Since $Sy_\nu = T(-y_\nu)$ and T is injective and has closed image, y_ν converges in E . Hence $U \cap \text{Ker}(S+T)$ is relatively compact. Since a locally compact Fréchet space is always finite-dimensional, $\dim \text{Ker}(S+T) < \infty$.

There exists a (closed) subspace E_1 of E such that $E = \text{Ker}(S+T) \oplus E_1$. Let $T_1 = T|E_1$ and $S_1 = S|E_1$. By replacing E by E_1 , T by T_1 , and S by S_1 , we can assume without loss of generality that $S+T$ is injective.

Suppose $\{x_\nu\} \subset E$ and $(S+T)x_\nu \longrightarrow z \in F$. Let

$$\lambda_\nu = \inf\{\lambda > 0 \,|\, x_\nu \in \lambda U\} .$$

We claim that there exists $A > 0$ such that $\lambda_\nu < A$ for all ν . Suppose the contrary. By replacing $\{x_\nu\}$ by a subsequence, we can assume that $\lambda_\nu \longrightarrow \infty$ and $\lambda_\nu > 0$. Let $y_\nu = \frac{x_\nu}{2\lambda_\nu}$. Then $y_\nu \in U$. By replacing $\{x_\nu\}$ by a subsequence if necessary, we can assume that Sy_ν converges in F. Hence $Ty_\nu = (S+T)y_\nu - Sy_\nu$ converges in F. Since T is injective and has closed image, y_ν converges in E. However, $y_\nu \notin \frac{1}{4}U$. Hence $y_\nu \not\longrightarrow 0$. On the other hand,

$$(S+T)y_\nu = \frac{1}{2\lambda_\nu}(S+T)x_\nu \longrightarrow 0,$$

because $(S+T)x_\nu \longrightarrow z$. This contradicts the injectivity of $S+T$. Hence $\lambda_\nu < A$ for all ν for some $A > 0$. Since $\frac{x_\nu}{A} \in U$, there exists a subsequence $\{u_\nu\}$ of $\{x_\nu\}$ such that $S\frac{u_\nu}{A}$ converges in F. It follows that

$$Tu_\nu = (S+T)u_\nu - AS\frac{u_\nu}{A}$$

converges in F. Consequently u_ν converges to some $x \in E$ and $(S+T)x = z$. Q.E.D.

(6.A.3) Proposition. Suppose $(X,\mathcal{O})$ is an n-dimensional complex manifold, Y is a relatively compact open subset of X, Ω is the sheaf of germs of holomorphic n-forms on X, and $\mathcal{F}$ is a locally free sheaf on X. Suppose, for some $p \geq 0$, both $H^p(X,\mathcal{F})$ and $H^p(Y,\mathcal{F})$ are Hausdorff and the map $H_c^{n-p}(Y, \mathcal{H}om_{\mathcal{O}}(\mathcal{F},\Omega)) \longrightarrow H_c^{n-p}(X, \mathcal{H}om_{\mathcal{O}}(\mathcal{F},\Omega))$ is surjective. Then $H^{p+1}(X,\mathcal{F})$ is Hausdorff.

Proof. Let $\mathfrak{U}$ be a countable Stein covering of X and $\mathfrak{V}$ is a countable Stein covering of Y such that every member of $\mathfrak{V}$ is relatively compact in some member of $\mathfrak{U}$. Consider the following commutative diagram:

$$\begin{array}{ccccc} C^{p-1}(\mathfrak{U},\mathcal{F}) & \xrightarrow{\alpha} & C^{p}(\mathfrak{U},\mathcal{F}) & \xrightarrow{\beta} & C^{p+1}(\mathfrak{U},\mathcal{F}) \\ \downarrow & & \downarrow \mu & & \downarrow \nu \\ C^{p-1}(\mathfrak{V},\mathcal{F}) & \xrightarrow{\sigma} & C^{p}(\mathfrak{V},\mathcal{F}) & \xrightarrow{\tau} & C^{p+1}(\mathfrak{V},\mathcal{F}) \end{array},$$

where the horizontal maps are coboundary maps and the vertical maps are restriction maps. We use * to denote the dual or the transpose. Taking duals and transposes in the above diagram, we obtain the following commutative diagram:

$$\begin{array}{ccccc} C^{p-1}(\mathfrak{U},\mathcal{F})^* & \xleftarrow{\alpha^*} & C^{p}(\mathfrak{U},\mathcal{F})^* & \xleftarrow{\beta^*} & C^{p+1}(\mathfrak{U},\mathcal{F})^* \\ \uparrow & & \uparrow \mu^* & & \uparrow \nu^* \\ C^{p-1}(\mathfrak{V},\mathcal{F})^* & \xleftarrow{\sigma^*} & C^{p}(\mathfrak{V},\mathcal{F})^* & \xleftarrow{\tau^*} & C^{p+1}(\mathfrak{V},\mathcal{F})^* \end{array}.$$

By (6.A.1), we have the following isomorphisms:

$$\operatorname{Ker} \alpha^*/\operatorname{Im} \beta^* \approx H_c^{n-p}(X, \mathcal{H}om_{\mathcal{O}}(\mathcal{F},\Omega)) ,$$
$$\operatorname{Ker} \sigma^*/\operatorname{Im} \tau^* \approx H_c^{n-p}(Y, \mathcal{H}om_{\mathcal{O}}(\mathcal{F},\Omega)) .$$

It follows that the map

$$\theta : C^{p+1}(\mathfrak{U},\mathcal{F})^* \oplus \operatorname{Ker} \sigma^* \longrightarrow \operatorname{Ker} \alpha^*$$

defined by

$$\theta(a\oplus b) = \beta^*(a) + \mu^*(b)$$

is surjective. θ^* is injective and has closed image. Let

$$\bar{\beta}: C^p(\mathfrak{U},\mathcal{F})/\mathrm{Im}\ \alpha \longrightarrow C^{p+1}(\mathfrak{U},\mathcal{F}) ,$$

$$\bar{\mu}: C^p(\mathfrak{U},\mathcal{F})/\mathrm{Im}\ \alpha \longrightarrow C^p(\mathcal{V},\mathcal{F})/\mathrm{Im}\ \sigma$$

be induced respectively by β and μ. Since $\mathrm{Im}\ \alpha$ and $\mathrm{Im}\ \sigma$ are closed, $C^p(\mathfrak{U},\mathcal{F})/\mathrm{Im}\ \alpha$ (respectively $C^p(\mathcal{V},\mathcal{F})/\mathrm{Im}\ \sigma$) is dual to $\mathrm{Ker}\ \alpha^*$ (respectively $\mathrm{Ker}\ \sigma^*$). Hence θ^* equals

$$\xi: C^p(\mathfrak{U},\mathcal{F})/\mathrm{Im}\ \alpha \longrightarrow C^{p+1}(\mathfrak{U},\mathcal{F}) \oplus \left(C^p(\mathcal{V},\mathcal{F})/\mathrm{Im}\ \sigma\right) ,$$

where $\xi(c) = \bar{\beta}(c) \oplus \bar{\mu}(c)$. Let

$$\eta : C^p(\mathfrak{U},\mathcal{F})/\mathrm{Im}\ \alpha \longrightarrow C^{p+1}(\mathfrak{U},\mathcal{F}) \oplus \left(C^p(\mathcal{V},\mathcal{F})/\mathrm{Im}\ \sigma\right)$$

be defined by $\eta(c) = -\bar{\mu}(c)$. Since η is compact, by (6.A.2), $\xi+\eta$ has closed image. It follows that $H^{p+1}(\mathfrak{U},\mathcal{F})$ is Hausdorff. Q.E.D.

CHAPTER 7

EXTENSION OF COHERENT ANALYTIC SHEAVES

This chapter is devoted to the proof of the general extension theorem for coherent sheaves.

(7.1) Proposition. *Suppose* $0 \leq a < b$ *in* $\mathbb{R}^N$, D *is a domain in* $\mathbb{C}^n$, $k \geq n$, *and* $\mathcal{F}$ *is a coherent analytic sheaf on* $D \times G^N(a,b)$ *such that* $\mathcal{F}^{[k]} = \mathcal{F}$ *and* $0_{[k+2]}\mathcal{F} = \mathcal{F}$. *Suppose either* $k > n$ *or there exists a thick set* A *in* D *such that, for* $t \in A$, $\mathcal{F}(t)$ *can be extended to a coherent analytic sheaf on* $\{t\} \times \Delta^N(b)$. *Then* $\mathcal{F}$ *can be extended to a coherent analytic sheaf* $\tilde{\mathcal{F}}$ *on* $D \times \Delta^N(b)$ *satisfying* $\tilde{\mathcal{F}}^{[k]} = \tilde{\mathcal{F}}$.

Proof. Let $V = \operatorname{Supp} \mathcal{F}$. Since $0_{[k+1]}\mathcal{F} = 0$ and $0_{[k+2]}\mathcal{F} = \mathcal{F}$, every nonempty branch of V has pure dimension $k+2$. By (2.18), V can be extended to a subvariety $\tilde{V}$ of $D \times \Delta^N(b)$ whose every nonempty branch has dimension $k+2$. In the case $k = n$ let A' be the set of all $t \in A$ such that, for every open neighborhood U of t in D, $U \cap A$ is thick in U; and, by replacing A by A', we can assume that $A = A'$.

(a) Consider first the special case where $\dim \tilde{V} \cap (\{t\} \times \Delta^N(b)) \leq k+2-n$ for $t \in D$. We are going to show that, for every $t \in D$ when $k > n$, and, for every $t \in A$ when $k = n$, there exists an open neighborhood U of t in D such that $\mathcal{F} | U \times G^N(a,b)$ can be extended to a coherent analytic sheaf on $U \times \Delta^N(b)$.

Fix $t^0 \in D$ when $k > n$, and fix $t^0 \in A$ when $k = n$.

If $\tilde{V} \cap (\{t^0\} \times \Delta^N(b)) = \emptyset$, then for $a < b' < b$ in $\mathbb{R}^N$ there exists an open neighborhood U of t^0 in D such that $\tilde{V} \cap (D \times \Delta^N(b')) = \emptyset$ and $\mathcal{F} = 0$ on $U \times G(a,b')$. So we can assume that $\tilde{V} \cap (\{t^0\} \times \Delta^N(b))$ has pure dimension $k+2-n$.

Let $\ell = k+2-n$. Take $a < a' < b' < b$. By applying the theorem on the existence of special analytic polyhedra (i.e. (2.A.17)) to $\tilde{V} \cap (\{t^0\} \times \Delta^N(b))$, we can find

(i) holomorphic functions $f_1, \ldots, f_\ell$ on $\mathbb{C}^{n+N}$,

(ii) a connected Stein open neighborhood U of t^0 in D ,

(iii) an open neighborhood B of $\overline{U \times \Delta^N(a')}$ in $U \times \Delta^N(b')$, and

(iv) $0 < \alpha < \beta$ in $\mathbb{R}$,

such that

(1) the holomorphic map $F: \mathbb{C}^{n+N} \longrightarrow \mathbb{C}^{n+\ell}$ defined by the coordinates of $\mathbb{C}^n$ and $f_1, \ldots, f_\ell$ makes $\tilde{V} \cap B \cap F^{-1}(U \times \Delta^\ell(\beta))$ an analytic cover over $U \times \Delta^\ell(\beta)$ and

(2) $\tilde{V} \cap (\overline{U \times \Delta^N(a')}) \subset F^{-1}(U \times \Delta^\ell(\alpha))$.

Let

$$X = \tilde{V} \cap B \cap F^{-1}(U \times G^\ell(\alpha,\beta))$$

and let

$$\pi: X \longrightarrow U \times G^\ell(\alpha,\beta)$$

be induced by F . Let $\mathcal{F}' = \mathcal{F}|X$. The zeroth direct image $\pi_*^0 \mathcal{F}'$ of $\mathcal{F}'$ under π is a coherent analytic sheaf on $U \times G^\ell(\alpha,\beta)$. Since $(\pi_*^0 \mathcal{F}')(t) = \pi_*^0(\mathcal{F}(t))$ for $t \in U$, it follows that, in the case $k = n$, $(\pi_*^0 \mathcal{F}')(t)$ can be extended to a coherent analytic sheaf on $\{t\} \times \Delta^{\ell}(\beta)$ for $t \in A \cap U$. Since $(\pi_*^0 \mathcal{F}')^{[k]} = \pi_*^0 \mathcal{F}'$, by (6.11) $\pi_*^0 \mathcal{F}'$ can be extended to a coherent analytic sheaf on $U \times \Delta^{\ell}(\beta)$. Hence $\Gamma(U \times G^{\ell}(\alpha,\beta), \pi_*^0 \mathcal{F}')$ generates $\pi_*^0 \mathcal{F}'$. It follows that $\Gamma(X, \mathcal{F})$ generates $\mathcal{F}|X$. After shrinking U and $G^{\ell}(\alpha,\beta)$, we can assume without loss of generality that there exists a sheaf-epimorphism $\varphi: \mathcal{O}^p \longrightarrow \mathcal{F}$ on X, where $\mathcal{O}$ is the structure sheaf of $\tilde{V}$.

Let $\tilde{X} = \tilde{V} \cap B \cap F^{-1}(U \times \Delta^{\ell}(\beta))$ and let $\tilde{\pi}: X \longrightarrow U \times \Delta^{\ell}(\beta)$ be induced by F. Let $\mathcal{G} = \tilde{\pi}_*^0(\mathcal{O}^p|\tilde{X})$. $\mathcal{G}$ is coherent on $U \times \Delta^{\ell}(\beta)$ and $\pi_*^0(\operatorname{Ker} \varphi)$ is a coherent analytic subsheaf of $\mathcal{G}|U \times G^{\ell}(\alpha,\beta)$. Since $0_{[k+1]\mathcal{F}} = 0$, $\pi_*^0(\operatorname{Ker} \varphi)_{[k+1]\mathcal{G}} = \pi_*^0(\operatorname{Ker} \varphi)$. By (4.6), $\pi_*^0(\operatorname{Ker} \varphi)$ can be extended to a coherent analytic subsheaf of $\mathcal{G}$ on $U \times \Delta^{\ell}(\beta)$. Hence some subset of $\Gamma(U \times \Delta^{\ell}(\beta), \mathcal{G})$ generates $\pi_*^0(\operatorname{Ker} \varphi)$. It follows that some subset Λ of $\Gamma(\tilde{X}, \mathcal{O}^p)$ generates $\operatorname{Ker} \varphi$. Let $\mathcal{A}$ be the subsheaf of $\mathcal{O}^p|\tilde{X}$ generated by Λ. Then $(\mathcal{O}^p/\mathcal{A})|\tilde{X}$ is a coherent analytic sheaf on $\tilde{X}$ extending $\mathcal{F}|X$.

Let $\mathcal{F}^*$ be the sheaf on $\tilde{V} \cap (U \times \Delta^N(b))$ which agrees with $\mathcal{O}^p/\mathcal{A}$ on X and agrees with $\mathcal{F}$ on

$$\tilde{V} \cap (U \times \Delta^N(b)) - \tilde{\pi}^{-1}(U \times \overline{\Delta^{\ell}(\alpha)}) .$$

$\mathcal{F}^*$ is coherent. Let $\mathcal{F}^{\#} = \left(\mathcal{F}^*/0_{[k+1]\mathcal{F}^*}\right)^{[k]}$. Since $\mathcal{F}^{\#}$ agrees with $\mathcal{F}$ on $U \times G^N(b',b)$, by (3.16), $\mathcal{F}^{\#}$ agrees with $\mathcal{F}$ on $U \times G^N(a,b)$.

(b) We continue to assume that $\dim \tilde{V} \cap (\{t\} \times \Delta^N(b)) \leq k+2-n$ for $t \in D$. Let D' be the largest open subset of D such that $\mathcal{F}|D' \times G^N(a,b)$ can be extended to a coherent analytic sheaf on $D' \times \Delta^N(b)$. It follows from (a) that $D' = D$ when $k > n$, and D' is a nonempty closed subset of D when $k = n$. Hence in both cases $D' = D$.

(c) Let $\sigma: \tilde{V} \longrightarrow D$ be induced by the natural projection $D \times \Delta^N(b) \longrightarrow D$. Let S be the topological closure of the set of points where the rank of σ is $< n$. Take $a < b' < b$ in $\mathbb{R}^N$. By (2.A.7), $\sigma(S \cap \overline{(D \times \Delta^N(b'))})$ is a closed thin set in D. Let $D' = D - \sigma(S \cap (D \times \Delta^N(b')))$. By (b), $\mathcal{F}|D' \times G^N(a,b)$ can be extended to a coherent analytic sheaf on $D' \times \Delta^N(b)$.

Let L be an arbitrary relatively compact open subset of D. By (2.2) there exists an N-dimensional plane T in $\mathbb{C}^n \times \mathbb{C}^N$ such that for some nonempty open subset Q of L and for some connected open neighborhood R of L in D we have

(i) $(Q+T) \cap (\mathbb{C}^n \times \Delta^N(b)) \subset D' \times \Delta^N(b)$

(ii) $L \times \Delta^N(b) \subset (R+T) \cap (\mathbb{C}^n \times \Delta^N(b)) \subset D \times \Delta^N(b)$

(iii) $\dim(x+T) \cap \tilde{V} \leq k+2-n$ for $x \in V$,

where

$$Q+T = \{x+y \mid x \in Q,\ y \in T\}$$

and

$$x+T = \{x+y \mid y \in T\}.$$

By (b), $\mathcal{F}|(R+T) \cap (\mathbb{C}^n \times G^N(a,b))$ can be extended to a coherent analytic sheaf $\mathcal{F}^*$ on $(R+T) \cap (\mathbb{C}^n \times \Delta^N(b))$. It follows that $\mathcal{F}^*|L \times \Delta^N(b)$ extends $\mathcal{F}|L \times G^N(a,b)$. The proposition follows from the arbitrariness of L. Q.E.D.

(7.2) Proposition. Suppose $0 < a < b$ in $\mathbb{R}^N$, D is a connected Stein open subset of $\mathbb{C}^n$ with $H^2(D,\mathbb{Z}) = 0$, and $\mathcal{F}$ is a coherent analytic sheaf on $D \times \Delta^N(b)$. Suppose $1 \leq \nu \leq N-1$ and

$$0 \longrightarrow {}_{n+N}\mathcal{O}^{q_{N-\nu-1}} \longrightarrow \cdots \longrightarrow {}_{n+N}\mathcal{O}^{q_0} \longrightarrow \mathcal{F} \longrightarrow 0$$

is an exact sequence of sheaf-homomorphisms on $D \times \Delta^N(b)$. Suppose $\xi \in H^\nu(D \times G^N(a,b),\mathcal{F})$. For $t \in D$, let $\xi_t \in H^\nu(\{t\} \times G^N(a,b),\mathcal{F}(t))$ be the image of ξ under the map $H^\nu(D \times G^N(a,b),\mathcal{F}) \longrightarrow H^\nu(\{t\} \times G^N(a,b),\mathcal{F}(t))$ induced by the natural map $\mathcal{F} \longrightarrow \mathcal{F}(t)$. Suppose A is a thick set in D and for $t \in A$ there exists a holomorphic function f_t on $\Delta(b_1)$ such that f_t is nowhere zero on $G^1(a_1,b_1)$ and $(f_t \circ \pi)\xi_t = 0$, where $\pi:\mathbb{C}^N \longrightarrow \mathbb{C}$ is defined by $\pi(z_1, \ldots, z_N) = z_1$. Then there exists a holomorphic function F on $D \times \Delta(b_1)$ such that F is nowhere zero on $D \times G^1(a_1,b_1)$ and $(F \circ \Pi)\xi = 0$, where $\Pi : D \times \mathbb{C}^N \longrightarrow D \times \mathbb{C}$ is defined by $\Pi(t,z) = (t,\pi(z))$.

Proof. Let $p:D \times \Delta^N(b) \longrightarrow D$ be the natural projection. By (3.11) there exists a thin set C in D such that $\mathcal{F}$ is p-flat on $(D-C) \times \Delta^N(b)$. By replacing A by $A-C$, we can assume without loss of generality that $\mathcal{F}$ is p-flat on $A \times \Delta^N(b)$.

By (3.2), $H^\mu(D \times G^N(a,b),{}_{n+N}\mathcal{O}) = 0$ and $H^\mu(\{t\} \times G^N(a,b),{}_N\mathcal{O}) = 0$ for $1 \leq \mu \leq N-2$ and $t \in D$. It follows that, since $\mathcal{F}$ is p-flat on $A \times \Delta^N(b)$, it suffices to prove the special case where $\mathcal{F} = {}_{n+N}\mathcal{O}$ and $\nu = N-1$. Let $H = G^1(a_1,b_1) \times \ldots \times G^1(a_N,b_N)$. ξ can be defined by a holomorphic function Θ on $D \times H$ whose Laurent series expansion in $z_1, \ldots, z_N$ has the form

$$\Theta = \sum_{\nu_1,\ldots,\nu_N \leq -1} \Theta_{\nu_1\ldots\nu_N} z_1^{\nu_1} \ldots z_N^{\nu_N} .$$

For $t \in A$, let Θ_t be the restriction of Θ to $\{t\} \times H$. $(f_t \circ \pi)\xi_t = 0$ means that $(f_t \circ \pi)\,\Theta_t$ can be extended to a holomorphic function on

$$\tilde{H} := \Delta(b_1) \times G^1(a_2,b_2) \times \ldots \times G^1(a_N,b_N) .$$

Hence, for $t \in A$, Θ_t can be extended to a meromorphic function on $\tilde{H}$ whose pole-set Z_t is of the form $P_t \times G^1(a_2,b_2) \times \ldots \times G^1(a_N,b_N)$, where P_t is at most a finite subset of $\overline{\Delta(a_1)}$. Since by (2.A.8) the set $A \times G^1(a_2,b_2) \times \ldots \times G^1(a_N,b_N)$ is thick in $D \times G^1(a_2,b_2) \times \ldots \times G^1(a_N,b_N)$, by (1.1), Θ can be extended to a meromorphic function $\tilde{\Theta}$ on $D \times H$.

Let Z be the pole-set of $\tilde{\Theta}$. Let Z' be the set of all $x \in Z$ such that

$$\dim_x Z \cap \pi^{-1}\pi(x) \geq N-1 .$$

By (2.A.5), Z' is a subvariety of Z. Since

$$Z' = (D \times \tilde{H}) \cap \pi^{-1}\pi(Z') ,$$

$\pi(Z')$ is a subvariety of $D \times \Delta(b_1)$. Since $Z \cap (D \times H) = \emptyset$, $\pi(Z') \subset D \times \overline{\Delta(a_1)}$. Let V be the n-dimensional component of $\pi(Z')$. Since D is Stein and $H^2(D,\mathbb{Z}) = 0$, there exists a holomorphic function F on $D \times \Delta(b_1)$ whose zero-set is V. We claim that $(F^k \circ \Pi)\xi = 0$ for k sufficiently large. It suffices to show that $(F^k \circ \Pi)\tilde{\Theta}$ is holomorphic on $D \times \tilde{H}$ for k sufficiently large.

Let W be the union of all branches of $\Pi(Z')$ of dimension $< n$. Let $\sigma: D \times \Delta(b_1) \longrightarrow D$ be the natural projection. Clearly $\{t\} \times P_t \subset \Pi(Z')$ for $t \in A$. It follows that

$$(*) \qquad \{t\} \times P_t \subset V \quad \text{for } t \in A - \sigma(W).$$

For $k \geq 0$ let A_k be the set of all $t \in A - \sigma(W)$ such that the restriction of $(F^k \circ \Pi)\tilde{\Theta}$ to $\{t\} \times \tilde{H}$ is holomorphic on $\{t\} \times \tilde{H}$. By (*),

$$\bigcup_{k=0}^{\infty} A_k = A - \sigma(W).$$

Since $A - \sigma(W)$ is thick in D, there exists some k such that A_k is thick in D. By (2.A.8) and (1.2), $(F^k \circ \Pi)\tilde{\Theta}$ is holomorphic on $D \times \tilde{H}$. Q.E.D.

(7.3) Proposition. Suppose $0 < a < b < \tilde{b}$ in $\mathbb{R}^N$, $D \subset\subset \tilde{D}$ are Stein open subsets of $\mathbb{C}^n$, and $\mathcal{F}$ is a coherent analytic sheaf on $\tilde{D} \times \Delta^N(\tilde{b})$. Suppose $1 \leq \nu \leq N-2$ and $\operatorname{codh} \mathcal{F} \geq n+\nu+2$ on $D \times G^N(a,b)$. Then there exists a coherent ideal sheaf $\mathcal{J}$ on $D \times \Delta^N(b)$ whose zero-set is contained in $S_{n+\nu+1}(\mathcal{F})$ such that $\Gamma(D \times \Delta(b), \mathcal{J}) H^\nu(D \times G(a,b), \mathcal{F}) = 0$.

Proof. By shrinking $\tilde{D}$ and $\tilde{b}$, we can assume without loss of generality that there exists an exact sequence of sheaf-homomorphisms

$$0 \longrightarrow \mathcal{G} \longrightarrow {}_{n+N}\mathcal{O}^{q_{N-\nu-3}} \longrightarrow \cdots \longrightarrow {}_{n+N}\mathcal{O}^{q_0} \longrightarrow \mathcal{F} \longrightarrow 0$$

on $\tilde{D} \times \Delta^N(\tilde{b})$. Since by (3.2) $H^{\mu}(D \times G^N(a,b), {}_{n+N}\mathcal{O}) = 0$ for $1 \leq \mu \leq N-2$, by replacing $\mathcal{F}$ by $\mathcal{G}$, we can assume without loss of generality that $\nu = N-2$ and $\mathcal{F}$ is locally free on $D \times G^N(a,b)$.

Let $\mathcal{F}^* = \mathcal{H}om_{{}_{n+N}\mathcal{O}}(\mathcal{F}, {}_{n+N}\mathcal{O})$ and $\mathcal{F}^{**} = \mathcal{H}om_{{}_{n+N}\mathcal{O}}(\mathcal{F}^*, {}_{n+N}\mathcal{O})$. By shrinking $\tilde{D}$ and $\tilde{b}$, we can assume without loss of generality that there exists an exact sequence of sheaf-homomorphisms

$$(*) \qquad {}_{n+N}\mathcal{O}^{p_{N-1}} \longrightarrow \cdots \longrightarrow {}_{n+N}\mathcal{O}^{p_0} \longrightarrow \mathcal{F}^* \longrightarrow 0$$

on $\tilde{D} \times \Delta^N(\tilde{b})$. Applying the functor $\mathcal{H}om_{{}_{n+N}\mathcal{O}}(\cdot, {}_{n+N}\ )$ to (*), we obtain a sequence

$$0 \longrightarrow \mathcal{F}^{**} \longrightarrow {}_{n+N}\mathcal{O}^{p_0} \longrightarrow \cdots$$

$$\longrightarrow {}_{n+N}\mathcal{O}^{p_{N-3}} \xrightarrow{\ \alpha\ } {}_{n+N}\mathcal{O}^{p_{N-2}} \xrightarrow{\ \beta\ } {}_{n+N}\mathcal{O}^{p_{N-1}}$$

which is exact on $D \times \Delta^N(b) - S_{n+N-1}(\mathcal{F})$. Since $\mathcal{F} = \mathcal{F}^{**}$ on $D \times G^N(a,b)$ and $H^{\mu}(D \times G^N(a,b), {}_{n+N}\mathcal{O}) = 0$ for $1 \leq \mu \leq N-2$, it follows that $H^{N-2}(D \times G^N(a,b), \mathcal{F})$ is isomorphic to $H^1(D \times G^N(a,b), \operatorname{Ker} \alpha)$.

From the exact sequence

$$0 \longrightarrow \operatorname{Ker} \alpha \longrightarrow {}_{n+N}\mathcal{O}^{p} \longrightarrow \operatorname{Im} \alpha \longrightarrow 0$$

we conclude that $H^1(D \times G^N(a,b), \operatorname{Ker} \alpha)$ is isomorphic to the cokernel of $\Gamma(D \times \Delta^N(b), \operatorname{Im}\alpha) \longrightarrow \Gamma(D \times G^N(a,b), \operatorname{Im}\alpha)$.

Let $\mathcal{J}$ be the maximum sheaf of ideals on $D \times \Delta^N(b)$ such that $\mathcal{J}(\operatorname{Ker}\beta) \subset \operatorname{Im}\alpha$ on $D \times \Delta^N(b)$, i.e. $\mathcal{J} = (\operatorname{Im}\alpha : \operatorname{Ker}\beta)_{n+N^{\mathcal{O}}}$ on $D \times \Delta^N(b)$. Since $\Gamma(D \times G^N(a,b), \operatorname{Im}\alpha)$ is isomorphic to $\Gamma(D \times \Delta^N(b), \operatorname{Ker}\beta)$, the ideal-sheaf $\mathcal{J}$ satisfies the requirement. Q.E.D.

(7.4) Theorem (Sheaf Extension). *Suppose* $0 \leq a < b$ *in* $\mathbb{R}^N$, D *is a domain in* $\mathbb{C}^n$, *and* $\mathcal{F}$ *is a coherent analytic sheaf on* $D \times G^N(a,b)$ *such that* $\mathcal{F}^{[n]} = \mathcal{F}$. *Suppose* A *is a thick set in* D *and, for* $t \in A$, $\mathcal{F}(t)$ *can be extended to a coherent analytic sheaf on* $\{t\} \times \Delta^N(b)$. *Then* $\mathcal{F}$ *can be extended uniquely to a coherent analytic sheaf* $\tilde{\mathcal{F}}$ *on* $D \times \Delta^N(b)$ *satisfying* $\tilde{\mathcal{F}}^{[n]} = \tilde{\mathcal{F}}$.

Proof. Let k be the largest integer such that $0_{[k]}\mathcal{F} = 0$. We use descending induction on k . The case $k \geq n+N$ is trivial.

Let $\mathcal{G} = 0_{[k+1]}\mathcal{F}$ and $\mathcal{R} = \mathcal{F}/\mathcal{G}$. By (3.18) there exists a thin set C_1 in D such that $\mathcal{G}(t) = 0_{[k+1-n]}\mathcal{F}(t)$ for $t \in D - C_1$. It follows that, for $t \in A - C_1$, both $\mathcal{G}(t)$ and $\mathcal{R}(t)$ can be extended to coherent analytic sheaves on $\{t\} \times \Delta^N(b)$.

We are going to prove that $\mathcal{G}$ can be extended to a coherent analytic sheaf $\tilde{\mathcal{G}}$ on $D \times \Delta^N(b)$ satisfying $\tilde{\mathcal{G}}^{[n]} = \tilde{\mathcal{G}}$. When $k = n+1$, this follows from (7.1) and the fact that $\mathcal{G}^{[k-1]} = \mathcal{G}$. Assume that $k > n+1$. Let $\mathcal{A} = \mathcal{G}^{[k-1]}$. By

(7.1), $\mathcal{A}$ can be extended to a coherent analytic sheaf $\tilde{\mathcal{A}}$ on $D \times \Delta^N(b)$ satisfying $\tilde{\mathcal{A}}^{[n]} = \tilde{\mathcal{A}}$. By (3.17) there exists a thin set C_2 in D such that $\tilde{\mathcal{A}}(t)^{[0]} = \tilde{\mathcal{A}}(t)$ on $\{t\} \times \Delta^N(b)$ for $t \in D - C_2$. By (3.14) and (3.15), for $t \in A - C_1 - C_2$, $\operatorname{Im}(\mathcal{G}(t) \longrightarrow \tilde{\mathcal{A}}(t))$ can be extended to a coherent analytic subsheaf of $\tilde{\mathcal{A}}(t)$ on $\{t\} \times \Delta^N(b)$. By (4.5), $\mathcal{G}$ can be extended to a coherent analytic subsheaf $\tilde{\mathcal{G}}$ of $\tilde{\mathcal{A}}$ on $D \times \Delta^N(b)$ satisfying $\tilde{\mathcal{G}}_{[n]\tilde{\mathcal{A}}} = \tilde{\mathcal{G}}$.

Let $V_1 = S_{n+1}(\tilde{\mathcal{G}})$. Since $\tilde{\mathcal{G}}^{[n]} = \tilde{\mathcal{G}}$, by (3.13), $\dim V_1 \leq n-1$. Let $V_2 = \operatorname{Supp}(\mathcal{R}^{[n]}/\mathcal{R})$. $\dim V_2 \leq n$. Let A' be the set of all $t \in A$ such that, for every open neighborhood U of t in D, $U \cap A$ is thick in U. Let $\pi : D \times \Delta^N(b) \longrightarrow D$ be the natural projection.

(a) Assume first that $\dim \pi^{-1}(t) \cap V_j \leq 0$ for $j = 1,2$ and $t \in A'$. We are going to prove that for $t \in A'$ there exists an open neighborhood U of t in D such that $\mathcal{F} | U \times G^N(a,b)$ can be extended to a coherent analytic sheaf on $U \times \Delta^N(b)$.

Fix $t^0 \in A'$. We can choose $a < a' < b' < b$ in $\mathbb{R}^N$ and an open ball B of radius $r > 0$ centered at t^0 such that $B \subset\subset D$ and $V_j \cap (B \times G^N(a',b')) = \emptyset$ $(j = 1,2)$. Since $\mathcal{R}(t)$ can be extended to a coherent analytic sheaf on $\{t\} \times \Delta^N(b)$ for $t \in A - C_1$, by induction hypothesis, $\mathcal{R} | B \times G^N(a',b')$ can be extended to a coherent analytic sheaf $\tilde{\mathcal{R}}$ on $B \times \Delta^N(b')$ satisfying $\tilde{\mathcal{R}}^{[n]} = \tilde{\mathcal{R}}$.

Let $V' = \pi(V_1 \cap (B \times \Delta^N(b')))$. V' is a subvariety of B of dimension $\leq n - 1$. Since $A' \cap B$ is thick in B, by

(2.2) there exists a 1-dimensional plane H of $\mathbb{C}^n$ passing through t^0 such that $A' \cap B \cap H - V' \neq \emptyset$ and $\dim H \cap V' \leq 0$. After an isometric linear transformation of the coordinates of $\mathbb{C}^n$, we can assume without loss of generality that

$$H = \{t_2 = t_2^0, \ldots, t_n = t_n^0\} .$$

Since $\dim H \cap V' \leq 0$, there exist $t_1^* \in \mathbb{C}$ and $|t_1^0 - t_1^*| < r' < \frac{r}{2}$ such that $\{t \in H \,\big|\, |t_1 - t_1^*| = r'\}$ intersects A' but is disjoint from $H \cap V'$. After the coordinates transformation $(t_1', \ldots, t_n') = (t_1 - t_1^*, t_2 - t_2^0, \ldots, t_n - t_n^0)$ in $\mathbb{C}^n$, we can assume without loss of generality that $(t_1^*, t_2^0, \ldots, t_n^0) = 0$. There exist $0 < \gamma$ in $\mathbb{R}^{n-1}$ and $0 < \alpha < r' < \beta$ such that $\Delta^{n-1}(\gamma) \times \Delta(\beta) \subset Q$ and $\Delta^{n-1}(\gamma) \times G^1(\alpha,\beta)$ is disjoint from V' but intersects A'.

Since $\operatorname{codh} \tilde{\mathcal{G}} \geq n+2$ on $\Delta^{n-1}(\gamma) \times G^1(\alpha,\beta) \times \Delta^N(b')$, there exists an exact sequence of sheaf-homomorphisms

$$0 \longrightarrow {}_{n+N}\mathcal{O}^{q_{N-2}} \longrightarrow \cdots \longrightarrow {}_{n+N}\mathcal{O}^{q_0} \longrightarrow \tilde{\mathcal{G}} \longrightarrow 0$$

on $\Delta^{n-1}(\gamma) \times G^1(\alpha,\beta) \times \Delta^N(b')$.

Since $\tilde{\mathcal{R}}^{[n]} = \tilde{\mathcal{R}}$ and $\mathcal{F}^{[n]} = \mathcal{F}$, by (3.11) and (3.17) there exists a thin set C_3 in B such that

(i) $\mathcal{R}$ is π-flat on $(B - C_3) \times G^N(a', b')$,

(ii) $\tilde{\mathcal{R}}(t)^{[0]} = \tilde{\mathcal{R}}(t)$ on $\{t\} \times \Delta^N(b')$ for $t \in B - C_3$,

(iii) $\mathcal{F}(t)^{[0]} = \mathcal{F}(t)$ on $\{t\} \times G^N(a,b)$ for $t \in B - C_3$.

Hence, for $t \in A \cap B - C_3$, $\mathcal{F}(t)$ can be extended to a coherent analytic sheaf $\mathcal{F}(t)^{\sim}$ on $\{t\} \times \Delta^N(b)$ satisfying $(\mathcal{F}(t)^{\sim})^{[0]} = \mathcal{F}(t)^{\sim}$.

Let $W = \Delta^{n-1}(\gamma) \times G^1(\alpha,\beta)$ and $U = \Delta^{n-1}(\gamma) \times \Delta(\beta)$.
For $t \in A \cap W - C_3$ the following commutative diagram with exact rows

$$\begin{array}{ccccccccc} 0 & \longrightarrow & \mathcal{G} & \longrightarrow & \mathcal{F} & \xrightarrow{\lambda} & \mathcal{R} & \longrightarrow & 0 \\ & & \downarrow & & \downarrow & & \downarrow & & \\ 0 & \longrightarrow & \mathcal{G}(t) & \longrightarrow & \mathcal{F}(t) & \xrightarrow{\lambda(t)} & \mathcal{R}(t) & \longrightarrow & 0 \end{array}$$

on $U \times G^N(a',b')$ yields the exactness of the first 3 rows of the following commutative diagram:

$$\begin{array}{ccccc} \Gamma(U \times G^N(a',b'), \mathcal{F}) & \xrightarrow{\chi} & \Gamma(U \times G^N(a',b'), \mathcal{R}) & \xrightarrow{\tilde{\rho}} & H^1(U \times G^N(a',b'), \mathcal{G}) \\ \downarrow & & \downarrow & & \downarrow \\ \Gamma(W \times G^N(a',b'), \mathcal{F}) & \longrightarrow & \Gamma(W \times G^N(a',b'), \mathcal{R}) & \xrightarrow{\rho} & H^1(W \times G^N(a',b'), \mathcal{G}) \\ \downarrow & & \downarrow & & \downarrow \\ \Gamma(\{t\} \times G^N(a',b'), \mathcal{F}(t)) & \longrightarrow & \Gamma(\{t\} \times G^N(a',b'), \mathcal{R}(t)) & \xrightarrow{\rho_t} & H^1(\{t\} \times G^N(a',b'), \mathcal{G}(t)) \\ \sigma_t \uparrow & & \tau_t \uparrow & & \\ \Gamma(\{t\} \times \Delta^N(b'), \mathcal{F}(t)^{\sim}) & \xrightarrow{\lambda(t)^*} & \Gamma(\{t\} \times \Delta^N(b'), \tilde{\mathcal{R}}(t)) & & \end{array}$$

where σ_t and τ_t are restriction maps and $\lambda(t)^*$ is induced by $\lambda(t)^{\sim} : \tilde{\mathcal{F}}(t) \longrightarrow \tilde{\mathcal{R}}(t)$ which is the unique extension of $\lambda(t)$. By (3.14), σ_t and τ_t are isomorphisms. Since

$$\operatorname{Supp} \operatorname{Coker} \lambda(t)^{\sim} \subset \{t\} \times \overline{\Delta^N(a')} ,$$

for every $1 \leq j \leq N$ and $t \in A \cap W - C_3$ there exists a holomorphic function $f_{t,j}$ on $\{t\} \times \Delta(b_j)$ such that $f_{t,j}$ is nowhere zero on $\{t\} \times G^1(a_j,b_j)$ and $(f_{t,j} \circ \pi_j)\operatorname{Coker} \lambda(t)^{\sim} = 0$

on $\{t\} \times \Delta^N(b)$, where $\pi_j : \mathbb{C}^N \longrightarrow \mathbb{C}$ is defined by $\pi_j(z_1, \ldots, z_N) = z_j$. Hence $(f_{t,j} \circ \pi_j)\mathrm{Im}\ \rho_t = 0$ for $1 \le j \le N$ and $t \in A \cap W - C_3$. It follows from (7.2) that for $1 \le j \le N$ there exists a holomorphic function F_j on $W \times \Delta(b_j)$ such that F_j is nowhere zero on $W \times G^1(a_j, b_j)$ and $(F_j \circ \Pi_j)\mathrm{Im}\ \rho = 0$ where $\Pi_j : W \times \mathbb{C}^N \longrightarrow W \times \mathbb{C}$ is defined by $\Pi_j(t,z) = (t, \pi_j(z))$.

Consider the following portion of the Mayer-Vietoris sequence of $\tilde{\mathcal{G}}$ on $\Omega := (U \times G^N(a',b')) \cup (W \times \Delta^N(b'))$

$$(*) \quad \begin{aligned} H^1(\Omega, \tilde{\mathcal{G}}) &\longrightarrow H^1(U \times G^N(a',b'), \tilde{\mathcal{G}}) \oplus H^1(W \times \Delta^N(b'), \tilde{\mathcal{G}}) \\ &\longrightarrow H^1(W \times G^N(a',b'), \tilde{\mathcal{G}}). \end{aligned}$$

Let $\theta : H^1(\Omega, \tilde{\mathcal{G}}) \longrightarrow H^1(U \times G^N(a',b'), \tilde{\mathcal{G}})$ be the restriction map. Since $(F_j \circ \Pi_j)\mathrm{Im}\ \rho = 0$ for $1 \le j \le N$, it follows from (*) that

$$(\dagger) \qquad (F_j \circ \Pi_j)\mathrm{Im}\ \tilde{\rho} \subset \mathrm{Im}\ \theta .$$

Since $\mathrm{codh}\,\tilde{\mathcal{G}} \ge n+2$ on Ω , by (7.3) there exists a coherent ideal-sheaf $\mathcal{J}$ on $U \times \Delta^N(b')$ such that the zero-set of $\mathcal{J}$ is contained in $S_{n+1}(\tilde{\mathcal{G}}) \cap (U \times \Delta^N(b'))$ and

$$\Gamma(U \times \Delta^N(b'), \mathcal{J}) H^1(\Omega, \tilde{\mathcal{G}}) = 0 .$$

It follows from (†) that

$$(F_j \circ \Pi_j)\Gamma(U \times \Delta^N(b'), \mathcal{J})\mathrm{Im}\ \tilde{\rho} = 0$$

for $1 \le j \le N$. Hence

$$(\#) \qquad (F_j \circ \Pi_j)\Gamma(U \times \Delta^N(b'), \mathcal{J})\Gamma(U \times G^N(a',b'), \mathcal{R}) \subset \mathrm{Im}\ \chi \quad \text{for } 1 \le j \le N .$$

We know that $\Gamma(U \times G^N(a',b'),\mathcal{R})$ generates $\mathcal{R}$ on $U \times G^N(a',b')$ and $\Gamma(U \times G^N(a',b'),\mathcal{G})$ generates $\mathcal{G}$ on $U \times G^N(a',b')$. Since for $x \in U \times G^N(a',b')$ there exist $u \in (F_j \circ \Pi_j)$ for some $1 \leq j \leq N$ and $v \in \Gamma(U \times \Delta^N(b'),\mathcal{J})$ such that uv is nonzero at x, it follows from (#) that $\Gamma(U \times G^N(a',b'),\mathcal{F})$ generates $\mathcal{F}$ on $U \times G^N(a',b')$. After shrinking U and $G^N(a',b')$, we can assume without loss of generality that there exists a sheaf-epimorphism $\varphi: {}_{n+N}\mathcal{O}^p \longrightarrow \mathcal{F}$ on $U \times G^N(a',b')$. Since $(\mathrm{Ker}\ \varphi)_{[k+1]_{n+N}\mathcal{O}^p} = \mathrm{Ker}\ \varphi$, by (4.6) $\mathrm{Ker}\ \varphi$ can be extended to a coherent analytic subsheaf $\mathcal{K}$ of ${}_{n+N}\mathcal{O}^p$ on $U \times \Delta^N(b')$. ${}_{n+N}\mathcal{O}^p/\mathcal{K}$ is a coherent analytic sheaf on $U \times \Delta^N(b')$ extending $\mathcal{F} | U \times G^N(a',b')$. Hence $\mathcal{F} | U \times G^N(a,b)$ can be extended to a coherent analytic sheaf on $U \times \Delta^N(b)$.

(b) Now we assume that $\dim \pi^{-1}(t) \cap V_j \leq 0$ for $j = 1,2$ and $t \in D$. Let D' be the largest open subset of D such that $\mathcal{F} | D' \times G^N(a,b)$ can be extended to a coherent analytic sheaf on $D' \times \Delta^N(b)$. By (a), D' is a nonempty closed subset of D. Hence $D' = D$.

(c) Let Z' be the topological closure of the set of points of V_2 where the rank of $\pi | V_2$ is $< n$. Take $a < a' < b' < b$ in $\mathbb{R}^N$ and let

$$Z = \pi\left(\left(Z' \cap \left(D \times \overline{G^N(a',b')}\right)\right) \cup \left(V_1 \cap \left(D \times \overline{\Delta^N(b')}\right)\right)\right).$$

By (2.A.7), Z is a closed thin set in D. Let $D' = D - Z$. By (b), $\mathcal{F} | D' \times G^N(a,b)$ can be extended to a coherent analytic sheaf on $D' \times \Delta^N(b)$.

Let L be an arbitrary relatively compact open subset

of D. By (2.2) there exists an N-dimensional plane T in $\mathbb{C}^n \times \mathbb{C}^N$ such that for some nonempty open subset Q of L and for some connected open neighborhood R of L in D we have

(i) $(Q+T) \cap (\mathbb{C}^n \times \Delta^N(b)) \subset D' \times \Delta^N(b)$,

(ii) $L \times \Delta^N(b) \subset (R+T) \cap (\mathbb{C}^n \times \Delta^N(b)) \subset D \times \Delta^N(b)$,

(iii) $\dim(x+T) \cap V_j \leq 0$ for $x \in V_j$ $(j = 1,2)$,

where

$$Q+T = \{x+y \mid x \in Q,\ y \in T\}$$

and

$$x+T = \{x+y \mid y \in T\} .$$

By (b), $\mathcal{F} \mid (R+T) \cap (\mathbb{C}^n \times G^N(a,b))$ can be extended to a coherent analytic sheaf $\mathcal{F}^*$ on $(R+T) \cap (\mathbb{C}^n \times \Delta^N(b))$. It follows that $\mathcal{F}^* \mid L \times \Delta^N(b)$ extends $\mathcal{F} \mid L \times G^N(a,b)$. The theorem now follows from the arbitrariness of L . Q.E.D.

REFERENCES

[1] H. Alexander, Continuing 1-dimensional analytic sets, Math. Ann. 191 (1971), 143-144.

[2] J. Becker, Continuing analytic sets across $\mathbb{R}^n$, Math. Ann. 195 (1972), 103-106.

[3] E. Bishop, Conditions for the analyticity of certain sets, Michigan Math. J. 11 (1964), 289-304.

[4] J. Frenkel, Cohomologie nonabélienne et espaces fibrés, Bull. Soc. Math. France 83 (1957), 135-218.

[5] J. Frisch and G. Guenot, Prolongement de faisceaux analytiques cohérents, Invent. Math. 7 (1969), 321-343.

[6] P. Griffiths, Two theorems on extensions of holomorphic mappings, Invent. Math. 14 (1971), 27-62.

[7] R. C. Gunning and H. Rossi, Analytic Functions of Several Complex Variables, Englewood Cliffs, N.J.: Prentice-Hall, 1965.

[8] F. R. Harvey, Extension of positive currents, Preprint 1971.

[9] A. Hurwitz and R. Courant, Vorlesungen über allgemeine Funktionentheorie und elliptische Funktionen (mit einem Anhang von H. Röhrl), 4. Auflage, Berlin-Heidelberg-New York: Springer-Verlag 1964.

[10] R. Kaufman, A theorem of Radó, Math. Ann. 169 (1967), 282.

[11] E. E. Levi, Studii sur punti singolari essenziali delle funzioni analitiche di due o piu variabili complesse, Annali di Mat. Pura ed Appl. 17, 3 (1910), 61-87.

[12] A. Markoe, Analytic families of differential complexes, J. Functional Analysis 9 (1972), 181-188.

[13] T. Radó, Über eine nicht-fortsetzbare Riemannsche Mannigfaltigkeit, Math. Zeitschr. 20 (1924), 1-6.

[14] R. Remmert and K. Stein, Über die wesentlichen Singularitäten analytischer Mengen, Math. Ann. 126 (1953), 263-306.

[15] W. Rothstein, Ein neuer Beweis des Hartogschen Hauptsätzes und seine Ausdehnung auf meromorphe Funktionen, Math. Zeitschr. 53 (1950), 84-95.

[16] ________, Zur Theorie der analytischen Mannifaltigkeiten im Raume von n komplexen Veränderlichen, Math. Ann. 129 (1955), 96-138.

[17] G. Scheja, Riemannsche Hebbarkeitssätze für Cohologieklassen, Math. Ann. 144 (1961), 345-360.

[18] E. Schmidt, Über den Millouxschen Satz, Sitzungsber. Preub. Akad. Wiss., Physik.-Math. Kl. (1932), 394-401.

[19] B. Schiffman, On the removal of singularities of analytic sets, Michigan Math. J. 15 (1968), 111-120.

[20] ________, Extension of positive line bundles and meromorphic maps, Invent. Math. 15 (1972), 332-347.

[21] Y.-T. Siu, Absolute gap-sheaves and extensions of coherent analytic sheaves, Trans. Amer. Math. Soc. 141 (1969), 361-376.

[22] __________, Extending coherent analytic sheaves, Ann. of Math. 90 (1969), 108-143.

[23] __________, An Osgood type extension theorem for coherent analytic sheaves, Proc. Conf. Several Complex Variables (College Park, Maryland 1970), Lecture Notes in Math. Vol 185, pp.189-241, Berlin-Heidelberg-New York: Springer Verlag 1971.

[24] __________, A Hartogs type extension theorem for coherent analytic sheaves, Ann. of Math. 93 (1971), 166-188.

[25] __________, A Thullen type extension theorem for positive holomorphic vector bundles, Bull. Amer. Math. Soc. 78 (1972), 775-776..

[26] __________, Analyticity of sets associated to Lelong numbers and the extension of meromorphic maps, Bull. Amer. Math. Soc. 79 (1973), 1200-1205.

[27] __________ and G. Trautmann, Extension of coherent analytic subsheaves, Math. Ann. 188 (1970), 128-142.

[28] __________ and __________, Gap-sheaves and the Extension of Coherent Analytic Subsheaves, Lecture Notes in Math. Vol. 172, Berlin-Heidelberg-New York: Springer Verlag 1971.

[29] K. Suominen, Duality for coherent sheaves on analytic manifolds, Ann. Acad. Sci. Fennical A. I. 424 (1968), 1-19.

[30] W. Thimm, Lückengarben von kohärenten analytischen Modulgarben, Math. Ann. 148 (1962), 372-394.

[31] ________, Fortsetzung von kohärenten analytischen Modulgarben, Math. Ann. 184 (1970), 329-353.

[32] P. Thullen, Über die wesentlichen Singularitäten analytischer Funktionen und Flächen im Raume von n Komplexen Veränderlichen, Math. Ann. 111 (1935), 137-157.

[33] G. Trautmann, Ein Kontinuitätssatz für die Fortsetzung kohärenter analytischer Garben, Arch. Math. 18 (1967), 188-196.

[34] ________, Cohérence de faisceaux analytiques de la cohomologie locale, C. R. Acad. Sci. Paris 267 (1968), 694-695.

INDEX

INDEX OF SYMBOLS

NOTES

NOTES

NOTES